Set Theory

CHARLES C. PINTER
Bucknell University

ADDISON-WESLEY PUBLISHING COMPANY
Reading, Massachusetts
Menlo Park, California · London · Amsterdam · Don Mills, Ontario · Sydney

This book is in the
ADDISON-WESLEY SERIES IN MATHEMATICS

Consulting Editor: LYNN H. LOOMIS

 Printed in the United States of America. Published simultaneously in Canada. Library of Congress Catalog Card No. 77-131203.

ISBN 0-201-05827-8
ABCDEFGHIJ-AL-79876

Preface

This book has been written for junior and senior students of mathematics; it is intended as a basic text for one-semester courses in set theory or the foundations of mathematics.

My chief concern throughout the writing of this work has been to make it accessible to relatively unsophisticated students. The arguments are perfectly rigorous, but I have attempted to make them as simple as possible, balancing the advantages of brevity with the need for sufficient detail. Every definition is accompanied by a fair amount of commentary, whose purpose is to motivate and explain new concepts in simple terms and provide some background information. There is a progressive increase in complexity which is carefully planned; the book begins with a repetition of the familiar arguments of elementary set theory. Then, throughout the first four chapters the level of abstract thinking gradually rises and the arguments become more difficult by small stages. This is designed so that even the less mature student may be brought to the point where he can understand the deeper results of set theory, which are presented, especially, in the last three chapters.

There is a vast literature of elementary textbooks in which set theory is treated from an intuitive point of view. This may be excused, and may even be desirable, at the high-school level; but there is no justification for teaching a theory which is known to be inconsistent to university mathematics majors. Set theory must be presented within the framework of an axiomatic system; however, as every lecturer who has taught such a course knows, it is difficult to choose an axiomatization of set theory which is both simple and easy to motivate for undergraduate students. The main disadvantage of the Zermelo-Fraenkel theory is that the major conceptual difficulties appear *at the very beginning*, when the student is least able to cope with them. For example, it needs to be proved that the union and Cartesian product of any two sets *exist*, that the domain and range of an arbitrary function exist, and so forth. On the other hand, although axiomatizations of the von Neumann-Gödel variety are more subtle and abstract, they *can* be made to appear fairly natural to the beginner (if carefully handled) and, most important of all, the chief conceptual difficulties can be pushed back to quite late in the course.

The present text begins with a brief account of the growth of set theory, with special emphasis on the problems which led to the development of the various systems of axiomatic set theory. These systems are compared, and their major advantages and disadvantages are studied. In the next chapter, the undefined notions and elementary axioms of set theory are introduced. The axiomatization used here is an adaptation of the system of von Neumann-Bernays-Gödel. All the objects of the theory are named *classes*; the student is instructed to think of a class as a "collection of objects" in the usual sense. However, for reasons which are carefully explained in the Introduction, a distinction must be made between classes and sets: every set is a class, but not all classes are sets. In particular, the reader is advised to think of sets as those collections which are not so all-inclusive as to lead to the "paradoxes." The particular properties of sets are not emphasized until later in the book; as a result, Chapter 1 contains a development of the algebra of classes which is hardly distinguishable from the development of intuitive set algebra to be found in an elementary text. The next three chapters, on functions, relations, and partially ordered classes, respectively, give a relatively orthodox treatment of these subjects; the fact that we call our objects *classes* rather than *sets* does not require us to alter in any way the standard development of these ideas.

To recapitulate, then, the axiomatic system used in this book makes it possible to present the elementary properties of classes, functions, relations, and ordering in a way which does not differ substantially from the common intuitive treatment of these topics. After this, the difference between classes and sets is re-affirmed, and the reader is reminded of the things which can be done with sets but cannot be done with classes except at the risk of producing "paradoxes." From this point on, the book deals chiefly with sets. The Axiom of Choice is introduced, and is shown to be equivalent to Zorn's Lemma and the well-ordering theorem. The natural numbers are presented, followed by a chapter devoted to the properties of infinite and denumerable sets. Then the transfinite cardinal and ordinal numbers are introduced.

It is fairly common practice, nowadays, to present transfinite arithmetic to upperclass students in an "abstract" fashion, roughly as follows: the class of the ordinal numbers is constructed, followed by a proof of the transfinite recursion theorem; operations on the ordinals are defined via transfinite recursion and the properties of these operations are then proved inductively; finally, the cardinal numbers are introduced as initial ordinals. While this approach is elegant and, in fact, eminently correct, it is in many ways unsound pedagogy, especially in a course for undergraduates. Indeed, the concept of cardinal and ordinal numbers is concrete, tangible, and easy to motivate; it seems a great pity to conceal the underlying simplicity of these objects in a formal system where they are treated, essentially, as abstract symbols.

In the present text, the cardinal numbers are introduced before the ordinals.

The concept of cardinal number is explained in the following way: the natural numbers, in their capacity as cardinals, serve as a set of standards for measuring the "size" of every finite set. A natural extension of this idea, the class CD of all the cardinal numbers serves as a collection of standards for measuring the "size" of *every* set. In Chapter 8, the class CD of the cardinals is defined to be a class with the following property: for every set A, there is a unique element $a \in \mathrm{CD}$ such that A and a are in one-to-one correspondence. The existence of the class of all the cardinals is assured by means of a provisional axiom (to be removed later) called the Axiom of Cardinality. The arithmetic of the cardinals is then presented using the properties of one-to-one correspondence.

The ordinal numbers are treated in a similar fashion. They are presented as a natural extension of the simple concept of finite ordinality. It is explained to the reader that the class OR of all the ordinal numbers serves as a collection of standards for designating the "position" of any element in any well-ordered set. In the beginning of Chapter 9 the class OR is defined to be a class with the following property: if A is any well-ordered set, there exists a unique $\alpha \in \mathrm{OR}$ such that A is similar to α; the existence of the class OR is guaranteed by means of a provisional axiom called the Axiom of Ordinality. The arithmetic of the ordinals is then presented using the properties of similarity of well-ordered sets.

In the last section of Chapter 9—a section which may be omitted by classes which are not sufficiently advanced—the class OR of the ordinal numbers is constructed using, essentially, the well-known method due to von Neumann. It is shown that the class of the initial ordinals may be used as the class CD of the cardinals. Finally, in Chapter 10, which may be entirely omitted by slower classes, the transfinite recursion theorem is proved, and it is shown that each of the operations on the ordinals has an equivalent definition using transfinite recursion. Several advanced topics are then presented, ending with a discussion of inaccessible ordinals and cardinals.

I am indebted to several anonymous reviewers for their valuable suggestions, which I incorporated into my manuscript. I am grateful to Addison-Wesley for their expert editing and their patience and help in getting this manuscript ready for production.

Lewisburg, Pennsylvania C.C.P.
November 1970

Contents

0

Historical Introduction

1 THE BACKGROUND OF SET THEORY

Although set theory is recognized to be the cornerstone of the "new" mathematics, there is nothing essentially new in the intuitive idea of a set. From the earliest times, mathematicians have been led to consider sets of objects of one kind or another, and the elementary notions of modern set theory are implicit in a great many classical arguments. However, it was not until the latter part of the nineteenth century, in the work of Georg Cantor (1845–1918), that sets came into their own as the principal object of a mathematical theory.

Strangely, it was his work in the highly technical field of trigonometric series which first led Cantor to study the properties of sets. At first, he confined himself to certain particular sets of real numbers which occurred in connection with the convergence of series. But Cantor was quick to understand that his discoveries applied to sets quite generally; in a series of remarkable papers, published between 1873 and 1897, he moved progressively further from the concrete problems which had initiated his thinking on sets, and toward the powerful general concepts which underlie set theory today.

The boldest step which Cantor had taken—in the eyes of his contemporaries—was his use of infinite sets, which he considered as no less natural than using finite sets. The question of "infinity" had long been one of the most sensitive problems of mathematics. The reader is undoubtedly acquainted with Zeno's famous "paradox," in which a unit line segment is divided into subintervals by the points 1/2, 1/4, 1/8, 1/16, etc. Each subinterval—no matter how small—has a definite, nonzero length, and there are infinitely many subintervals; hence, the seemingly paradoxical conclusion that infinitely many nonzero lengths can be added together to produce a finite length. In order to avoid such traps, classical mathematicians made a distinction between the "actual" infinite—in which infinitely many objects are conceived of as existing simultaneously—and the "virtual" infinite, which is simply the potential to exceed any given finite quantity. The "virtual" infinite was regarded as safe, hence admissible, whereas the "actual" infinite was taboo.

It is not surprising then, that Cantor's theory—with its uninhibited use of infinite sets (the notion of infinite was obviously understood here in the "actual" sense)—was not immediately accepted by his contemporaries. It was received at first with skepticism, sometimes even with open hostility. However, by the 1890's the more "palatable" parts of Cantor's theory were widely used, for they provided an elegant framework for a wide variety of mathematical theories. And before the turn of the century, even the most revolutionary aspects of set theory had been accepted by a great many mathematicians—chiefly because they turned out to be invaluable tools, particularly in analysis.

Meanwhile, the work of several outstanding mathematicians, in particular Dedekind, was taking a turn which would cast set theory in its most promising role—as the fundamental, "unifying" branch of mathematics. From the earliest times, mathematicians have given thought to the possibility of unifying the entire discipline under a small number of basic principles. Many of the ancient schools, from Euclid through the Middle Ages, contended that the various branches of mathematics could be subsumed under geometry (numbers might be conceived as geometric proportions); a far more successful attempt at unification came in the nineteenth century, when the work of Weierstrass, Dedekind, and others suggested that all of classical mathematics could be derived from the arithmetic of the natural numbers (positive integers). It was shown that every real number can be regarded as a sequence (called a "Cauchy sequence") of rational numbers; hence the study of the real numbers is reduced to that of the rational numbers. But the rational numbers can easily be regarded as pairs of integers, so finally the mathematics of the real numbers—which includes the calculus and (via analytic geometry) all of geometry—can be based on the natural numbers.

It was at this crucial point in the evolution of ideas on the foundations of mathematics that Dedekind, in his little book *Was sind und was sollen die Zahlen* (1888), revealed that the concept of natural numbers can be derived from the basic principles of set theory. A modern way to show this is the following: we let "0" be the empty set (that is, the set with no elements, denoted by the symbol $\varnothing$); "1" is defined to be the set $\{\varnothing\}$, that is, the set (of sets) containing the one element $\varnothing$. Then, "2" is defined to be the set $\{0, 1\}$, "3" is defined to be $\{0, 1, 2\}$, and so on. All the properties of the natural numbers can be proven using these definitions and elementary set theory.

By the turn of the century, then, set theory had not only been accepted as an indispensable tool by a large segment of the mathematical community, but, moreover, it was a serious contender for the position of primacy among the mathematical sciences.

Ironically, at the very time when Cantor's ideas seemed finally to have gained acceptance, the first of certain "paradoxes" were announced, which eventually cast serious doubts as to the basic soundness of set theory in its "Cantorian" form. These paradoxes had such wide repercussions that it is worth looking at them in some detail.

2 THE PARADOXES

Between 1895 and 1910 a number of contradictions were discovered in various parts of set theory. At first, mathematicians paid little attention to them; they were termed "paradoxes" and regarded as little more than mathematical curios. The earliest of the paradoxes was published in 1897 by Burali-Forti, but it had already been discovered, two years earlier, by Cantor himself. Since the Burali-Forti paradox appeared in a rather technical region of set theory, it was hoped, at first, that a slight alteration of the basic definitions would be sufficient to correct it. However, in 1902 Bertrand Russell gave a version of the paradox which involved the most elementary aspects of set theory, and therefore could not be ignored. In the ensuing years other contradictions were discovered, which seemed to challenge many of the "safest" notions of mathematics.

The "paradoxes" of set theory are of two different kinds, the one called *logical* paradoxes, the other called *semantic* paradoxes. The reason for the names "logical" and "semantic" will become clear to us when we have seen a few examples of these paradoxes; essentially, the "logical" paradoxes arise from faulty logic whereas the "semantic" paradoxes arise from the faulty use of language.

We will devote the remainder of this section to the presentation of two of the most celebrated paradoxes, which involve only elementary concepts of set theory. The first is a "logical" paradox, the second is a "semantic" paradox; both may be considered as typical of their kind.

The simplest of the logical paradoxes is *Russell's paradox*, which can be described as follows:

> If A is a set, its elements may themselves be sets; this situation occurs frequently in mathematics—for example, A may be a set of lines, where each line is regarded as a set of points. Now the possibility arises that A may be an element of itself; for example, the set of all sets has this property.
>
> Let S denote the *set of all sets that are not elements of themselves.* Is S an element of itself? Well, if S *is* an element of S, then—by the very definition of S—S is *not* an element of S. If S is *not* an element of S, then (again, because of the way S is defined) S *is* an element of S. Thus, we have proven that S is an element of S if and only if S is not an element S—a contradiction of the most fundamental sort.

Usually, in mathematics, when we reach a contradiction of this kind, we are forced to admit that one of our assumptions was in error. In this case, we are led to conclude either that it is meaningless to speak of a set as being an element of itself, or that there is no such thing as a "set of all sets which are not elements of themselves." We will return to this question presently; meanwhile, let us say a few words about the semantic paradoxes.

Typical of the semantic paradoxes is *Berry's paradox*:

> For the sake of argument, let us admit that all the words of the English language are listed in some standard dictionary. Let T be *the set of all the natural numbers that can be described in fewer than twenty words of the English language.* Since there are only a finite number of English words, there are only finitely many combinations of fewer than twenty such words—that is, T is a finite set. Quite obviously, then, there are natural numbers which are greater than all the elements of T; hence there is a *least natural number which cannot be described in fewer than twenty words of the English language.* By definition, this number is not in T; yet we have described it in sixteen words, hence it is in T.

Once again, we are faced with a glaring contradiction; since the above argument would be unimpeachable if we admitted the existence of the set T, we are irrevocably led to the conclusion that a set such as T simply cannot exist.

Before the paradoxes, the question of the *existence* of sets had never been posed. Cantor "defined" a set to be "a collection of definite, distinguishable objects of our perception which can be conceived as a whole." More specifically, Cantor and his early followers accepted the "common-sense" notion that if we can describe a property of objects, we can also speak of the set of all objects possessing that property. The paradoxes had the singular merit of proving this naive conception of sets to be unacceptable—if only because certain "properties" lead to paradoxical sets.

In the various movements which sprang up, during the early 1900's, with the aim of revising the foundations of set theory, the topic of central concern was the *existence* of sets. What properties legitimately define sets? Under what conditions do properties define sets at all? How can new sets be formed from existing ones?

3 THE AXIOMATIC METHOD

The appearance of the paradoxes marked the beginning of a crisis in the foundations of mathematics which has not been completely resolved to our day. It became abundantly clear that the intuitive conception of a set, as embodied in Cantor's "definition," does not provide a satisfactory basis for set theory—much less for mathematics as a whole. Minor attempts to eliminate the paradoxes by excluding specific types of concepts and definitions were doomed to failure; nothing less than an entirely new approach was needed. Starting about 1905, several ways of dealing with the problem were proposed and developed by their adherents; most of them can be classified into three major groups, called the "axiomatic," the "logistic," and the "intuitionist" schools. The remainder of this chapter is devoted to presenting these three ways of thought. First, however, we shall briefly review the development of the axiomatic method.

The axiomatic method in mathematics emerged in a highly developed form, about 300 B.C., with the appearance of Euclid's *Elements.* Although the method popularized by Euclid has become a characteristic feature of every branch of mathematics today, only in comparatively recent years has it been applied outside of geometry. For this reason, our modern understanding of axiom systems, and of deductive reasoning generally, has to a great extent come out of studies in the field of geometry. It is worth examining a few of the major developments in geometry which influenced the growth of the axiomatic method.

To Euclid and his times, the axioms and postulates represent "truths" whose validity is beyond question. For example, it was this belief in the absolute truth of geometric propositions which led to the millenia-long controversy over Euclid's "parallel postulate." This postulate asserts that if two lines, A and B, intersect a third line C, and if the interior angles which A and B make with C (on a given side of C) add up to less than two right angles, then A and B necessarily intersect. Because this statement appeared to be "obviously true"—yet it lacked the terse simplicity of the other axioms and postulates—geometers from Euclid to the 1700's succeeded one another in vain attempts to prove it from the remaining assumptions. Only in the mid-nineteenth century was the question resolved when Bolyai and Lobachevski, each replacing the parallel postulate by an assumption which *contradicted* it, developed "non-Euclidean" geometries. The non-Euclidean geometries were shown to be no less consistent than Euclidean geometry, since they could be given Euclidean interpretations (that is, by suitably reinterpreting "point," "line," "angle," and so forth, the postulates of either Bolyai or Lobachevski can be made to hold in Euclidean geometry). Thus, not only is the parallel postulate independent of the other axioms and postulates of Euclid's system, but alternative, equally consistent geometries can be founded which do not describe the space of our everyday experience. With this came the recognition that axioms are not "universal truths," but are whatever statements we wish to use as premises in an argument.

Perhaps the greatest defect in the *Elements* is the number of tacit assumptions made by Euclid—assumptions not granted by the postulates. For example, in a certain proof it is assumed that two circles, each passing through the center of the other, have a pair of points in common—yet the postulates do not provide for the existence of these points. Elsewhere, Euclid speaks of a point as being *between* two others, yet he does not define "betweenness" or postulate any of its properties. Other arguments in the *Elements* involve the concept of rigid motion—a concept which is not defined or mentioned in the postulates.

Thus, throughout Euclid, the orderly chain of logical inferences is frequently broken by tacit appeals to visual evidence. With the discovery of these gaps, mainly in the nineteenth century, grew the understanding that a mathematical argument must be able to proceed without the mediation of spatial or other intuition; that certain objects and relations (such as "point," "line," "betweenness") must be regarded as *undefined notions* and their properties fully specified; that deduction is, in a very essential manner, independent of the *meaning* of concepts. In 1882, M. Pasch published the first formulation of geometry in which the exclusion of any appeal to intuition is clearly stated as a goal and systematically carried out.

By the end of the nineteenth century, then, a modern conception of the axiomatic method began to emerge. In its broad outlines, it did not differ from the ideas held by Euclid: a mathematical theory is "axiomatic" if certain statements are selected to be "axioms," and all the remaining propositions of the theory are derived from the axioms by logical inference. However, there was a new understanding of the *formal* nature of mathematical proof. Inasmuch as possible, the axioms should be sufficiently detailed, and the rules of logical deduction sufficiently explicit, that neither intuition nor intelligence is needed to go through the steps of a proof. Ideally, it should be possible for a computer to verify whether or not a proof is correct.

As long as mathematics is formulated in ordinary languages, such as English, human understanding is indispensable for interpreting statements and finding the structure of complex sentences. Thus, if intuition is to be completely removed from mathematical proof, an essential prerequisite is the development of a *formal* mathematical language: the "rules" of this language must be strictly codified, so that every statement is unambiguous and its structure clear. The creation of formal, symbolic languages was one of the most important developments of modern mathematics; here is what such a language looks like.

The most basic mathematical statements look like this:

"X is parallel to Y,"
"y lies between x and z,"
"X is an open set," etc.

They are statements about an object, a pair of objects, or more generally, about an ordered n-tuple of objects. These statements are called *elementary predicates*, and the letters X, Y, x, y, z are called their *variables*.

It is convenient to denote a predicate by a single letter followed by the list of its variables. Thus, "X is parallel to Y" may be written $A(X, Y)$, "y lies between x and z" may be written $B(x, y, z)$, and so on. Now the mature student is aware of the fact that the "meaning" of a predicate is immaterial in the process of mathematical reasoning. For example, the "meaning" of the word *parallel* has no bearing on the course of a geometrical argument; all that matters is the relationship between the statement "X is parallel to Y" and other statements such as "X intersects V" and "Y is perpendicular to Z." For this reason, elementary predicates are also called *atomic formulas*; they are integral, "indivisible," not to be analyzed further; they are only to be distinguished from each other.

It is a remarkable fact that every known branch of mathematics requires only a finite number (usually a very small number) of distinct elementary predicates. For example, every statement of plane Euclidean geometry can be expressed in terms of the following basic predicates:

1)

$P(x)$:	x is a point.
$L(x)$:	x is a line.
$B(x, y, z)$:	y is between x and z.
$E(x, y)$:	x equals y.
$I(x, y)$:	x belongs to y.
$C(u, v, x, y)$:	the segment uv is congruent to the segment xy.
$D(u, v, w, x, y, z)$:	the angle uvw is congruent to the angle xyz.

Set theory, as we shall see, may be formulated entirely in terms of the one predicate $x \in A$ (x is an element of A).

Predicates alone are not sufficient to express all the statements of mathematics, just as nouns alone would be inadequate to write English sentences. For example, we may wish to say that *if* "x is parallel to y" *and* "y is perpendicular to z," *then* "x is perpendicular to z." Such statements consist of predicates joined together by means of *logical connectives.* Thus, if P and Q are statements in our language, then so are the following:

$\neg P$:	not P.
$P \wedge Q$:	P and Q.
$P \vee Q$:	P or Q.
$P \Rightarrow Q$:	P implies Q.
$P \Leftrightarrow Q$:	P if and only if Q.

Finally, we may wish to say, for example, that if "x is a point" and "y is a point," then *there exists* a point z such that "z is between x and y." This requires the use of quantifiers. Thus, if $P(x)$ is a statement with a variable x, then the following are also statements:

$\forall x, P(x)$:	for every x, $P(x)$.
$\exists x \ni P(x)$:	there exists an x such that $P(x)$.

This completes our formal mathematical language. All of known mathematics can be expressed in terms of elementary predicates, logical connectives, and quantifiers. To illustrate how this language is used, let us take a simple example. The sentence

"If x and y are distinct points, then there is a point z between x and y"

can be symbolized as

2) $$[P(x) \wedge P(y) \wedge \neg E(x, y)] \Rightarrow [\exists z \ni (P(z) \wedge B(x, z, y))],$$

where the meaning of the predicates is given in (1) above.

One of the many benefits to be derived from the use of a formal language is that it is possible to describe precisely and explicitly the process of deduction in this language. A few clear, unambiguous rules decide when a statement T may be inferred from a statement S. A few such rules are the following:

3)
Rule A: from P and $P \rightarrow Q$ we may infer Q.
Rule B: from P and Q we may infer $P \wedge Q$.
Rule C: from $\neg(\neg P)$ we may infer P.
Rule D: from $P(c)$ we may infer $\exists x \ni P(x)$.

These, and a few other laws,* are called "rules of inference" in our language. A *formal* argument, from given premises, is a sequence of expressions of the formal language, where each expression is either a premise, or is derived from a preceding expression (or expressions) by applying one of the rules of inference.

Example Consider a formal language with a predicate $L(x, y)$; let us agree to write $x < y$ for $L(x, y)$. The following is a very simple formal argument in this language.

Premises

i) $a < b$.
ii) $b < c$.
iii) $[(a < b) \wedge (b < c)] \rightarrow (a < c)$.

Theorem $\exists x \ni [(a < x) \wedge (b < x)]$.

Proof.

Step	*Expression*	*Justification*
1	$a < b$	Premise (i)
2	$b < c$	Premise (ii)
3	$(a < b) \wedge (b < c)$	Steps 1, 2, and Rule B
4	$[(a < b) \wedge (b < c)] \rightarrow (a < c)$	Premise (iii)
5	$a < c$	Steps 3, 4, and Rule A
6	$(a < c) \wedge (b < c)$	Steps 5, 2, and Rule B
7	$\exists x \ni [(a < x) \wedge (b < x)]$	Step 6 and Rule D

* The four rules given in (3) above, together with seven additional rules, make up the system of natural deduction described in Slupecki and Borkowski [8]; these eleven rules are sufficient for every valid logical argument. Other systems are described in Quine [6] and Suppes [9].

The reader should note that the rules of inference are applied to expressions in a perfectly mechanical way. To all intents and purposes, the expressions can be regarded as meaningless arrays of symbols; the fact that they have a meaning *to us* is irrelevant to the task of carrying out the proof. Thus intuition is totally absent from a formal mathematical proof.

An axiomatic theory is said to be *formalized* if its axioms are transcribed in the formal language (for example, formula (2) on p. 8 is an axiom of Hilbert's plane geometry), and all of its proofs are formal proofs. While it is commonly accepted as the ideal, today, that every axiomatic theory be developed formally, it would be far too tedious, in practice, to do so. Symbolic statements are difficult to decipher, and formal proofs tend to be exceedingly long. Thus *mathematicians are usually content to satisfy themselves that an axiomatic theory can be formalized, and then proceed to develop it in an informal manner.* This will be our procedure in this book.

4 AXIOMATIC SET THEORY

To a great many mathematicians in the early 1900's, the answer to the problem posed by the paradoxes was to provide set theory with an axiomatic basis. The term "set" and the relation "is an element of" would be the undefined notions of such a theory, just as "point" and "line" are undefined notions in geometry; their "meaning" would be irrelevant, and their properties would be given formally by the axioms. In particular, *the axioms would be chosen in such a manner that all the useful results of Cantor's theory could be proven, whereas the paradoxes could not.*

The first axiomatization of set theory was given in 1908 by Zermelo. Zermelo's system, with certain modifications due to Skolem and Fraenkel, is widely used up to the present day. Zermelo wrote his work before the time when formal methods became widely understood and accepted; thus his set theory is not written in a formal language, but is closer in style to the older axiomatic treatments of geometry.

In Zermelo's system, there is one primitive relation, denoted by the symbol $\in$; the expression $x \in Y$ is to be read "x is an element of Y." The variables x, y, z, X, Y, etc., which we place to the right or to the left of the symbol $\in$, stand for objects which we agree to call "sets."

The reader may feel there ought to be *two* kinds of objects, namely sets and elements. Actually, this distinction is unnecessary: For, on the one hand, the relationship between element and set is a relative one rather than an absolute one (in fact, the element-set relationship is precisely the relation $\in$). On the other hand, almost every set in mathematics is a set *of sets.* For example, in plane analytic geometry, a line is a set of points; a point is a pair of real numbers (its coordinates); a real number is regarded as a sequence (that is, a set) of rational numbers; etc. Thus a useful simplification which is made in axiomatic set theory is to regard the elements of every set to be sets themselves; in other words, every set is considered to be a set *of sets*. This simplification has no harmful effects, and has the merit of reducing the number of primitive notions and axioms of set theory.

This suggests a comment on notation. Although it is customary to use small and capital letters as in $x \in Y$, it is in no way necessary. In fact, we will sometimes write things like $x \in y$ and $X \in Y$. All the variables in these expressions denote sets.

Almost every set that arises in our thinking is a set consisting of all the objects of a specified kind—that is, consisting of all the objects which satisfy a given condition. This is the most natural way in which sets occur: we are able to describe a condition on x—let us symbolize this condition by $S(x)$—and we are led to speak of the set of all objects x which satisfy $S(x)$.

Examples

The set of all objects x which satisfy the condition "x is an irrational number and $0 \leqslant x \leqslant 1$" (loosely speaking, the set of all irrational numbers between 0 and 1).

The set of all objects x which can be described by the sentence "x is a man" (loosely speaking, the set of all men).

Since this is the most natural way that sets arise, it is clearly desirable to have a principle in set theory which makes it possible—given any condition $S(x)$—to form the set of all objects x which satisfy $S(x)$. However, as we noted in Section 2, if such a principle is adopted *without any restrictions*, we are led to the paradoxes (for example, we can form the set of all sets which are not elements of themselves). Thus, we must devise such restrictions on this principle as will eliminate the paradoxes. Zermelo conceived the following restriction: Let $S(x)$ be a condition on x; we *cannot* form the set of all x which satisfy $S(x)$; but, if A is a given set, we can form the *set of all x in A which satisfy* $S(x)$. Thus, roughly speaking, a property of objects cannot be used to form a "new" set, but only to "select," from a set A whose existence has already been secured, all the elements which satisfy the given property.

Zermelo introduced this principle as an axiom in his system. Because its role is to *select* elements in sets, he called it the *axiom of selection* and stated it as follows:

> Let A be a set, and let $S(x)$ be a statement about x which is meaningful for every object x in A. There exists a set which consists of exactly those elements x in A which satisfy $S(x)$.

The set whose existence is given by the axiom of selection is customarily denoted by

$$\{x \in A \mid S(x)\}$$

[to be read: "the set of all x in A such that $S(x)$"]. Thus the reader should note that Zermelo's system does not allow us to form $\{x \mid S(x)\}$, [the "set of all x which satisfy $S(x)$"]; but, for any set A, we can form $\{x \in A \mid S(x)\}$.

How does the axiom of selection avoid the paradoxes? First, let us see what happens to Russell's paradox: the crucial set in Russell's argument is the "set of all sets which are not elements of themselves," which can be symbolized as $\{x \mid x \notin x\}$. As we have noted, this set cannot be formed in Zermelo's system; the best we can do is to produce $\{x \in A \mid x \notin x\}$, where A is any set which can be shown to exist. If we substitute $\{x \in A \mid x \notin x\}$ for $\{x \mid x \notin x\}$ in Russell's argument, the outcome changes completely. Indeed, let us go through the steps of the argument, with S denoting $\{x \in A \mid x \notin x\}$:

> $S \in S$ is impossible, for $S \in S$ implies $S \notin S$, a contradiction! Thus $S \notin S$. It follows that $S \notin A$, for if S were in A, then (because $S \notin S$) we would have $S \in S$, which would be a contradiction.

Hence Russell's argument merely proves that if A is any set, then the set $\{x \in A \mid x \notin x\}$ cannot be an element of A.

The other logical paradoxes disappear in similar fashion. The crucial sets in all of the logical paradoxes have a common trait: they are overly comprehensive—that is, they are "too large," they include too much. In Russell's paradox it is the "set of all sets which are not elements of themselves"; in Cantor's paradox (which is closely related to that of Russell) it is the "set of all sets." The axiom of selection cannot contribute to the formation of these "excessively large" sets, since it can only be used to form *subsets of existing sets.*

The problem of avoiding the semantic paradoxes is a more difficult one. The crucial sets in paradoxes such as Berry's are not "too large." The trouble seems, rather, to be inherent in the condition $S(x)$ which determines the set; even the restriction imposed by the axiom of selection is not an effective barrier. Thus, if $S(x)$ designates the sentence "x can be described in fewer than twenty words of the English language," then the offending set in Berry's paradox is $\{x \in N \mid S(x)\}$, where N denotes the set of the natural numbers. This set *can* be formed in Zermelo's system, if we admit $S(x)$ as an acceptable condition on x.

Thus, to prevent the semantic paradoxes, we must place restrictions on the type of "conditions" $S(x)$ which are admissible for determining sets. Zermelo attempted to do this by stipulating, in the axiom of selection, that $\{x \in A \mid S(x)\}$ can be formed only if *$S(x)$ is meaningful for every element x in A.* However, in so doing, he only raised new questions: How are we to understand "meaningful"? How do we determine whether $S(x)$ is meaningful?

We are forced, at last, to face a question which the alert reader may already have asked himself: What do we mean by a "condition" $S(x)$, by a "statement about an object x"? We cannot be content to regard the concept of a "statement about x" as intuitively known, since our purpose now is to axiomatize set theory, that is, to free it of all dependence on intuition. Zermelo failed to give a satisfactory answer to this question, because he did not frame his system in a formal language. However, in 1922, Skolem and Fraenkel, both working on formal axiomatizations of set theory, saw the natural way out of the dilemma: a "statement about x" is simply a statement *in the formal language* with one "free" variable x. (We say that x is *free* in $S(x)$ if x is not governed by a quantifier $\exists x$ or $\forall x$; thus, in $\exists y \ni (x < y)$, x is free whereas y is not).

In Zermelo's system there is only one elementary predicate, denoted by the symbol $\in$. Thus a statement in the formal language is an expression which can be written using only predicates $x \in Y$, $u \in V$, etc., logical connectives, and quantifiers.

If we restrict the "statements" $S(x)$ which can be used in the axiom of selection to those which are expressible in the formal language, we immediately eliminate all the semantic paradoxes. For example, there is no way of writing the sentence "x can be described in fewer than twenty words of the English language" in terms of the formal language. This solution—this way of avoiding the semantic paradoxes—is acceptable from the mathematical point of view, though it is hardly an ideal solution philosophically. Mathematically, we can still form all the sets essential for mathematics; from a broader point of view, though, we cannot form anything like the "set of all men," the "set of all Latin verbs," etc. No better solution has been devised to this day.

We have seen how Zermelo's system, with modifications due to Skolem and Fraenkel, manages to avoid the paradoxes. The remaining axioms of Zermelo's system are similar to those which will be developed in the following chapters. Essentially, they provide for the existence of the set of all the natural numbers (from which we can construct the other number systems of mathematics), and guarantee the existence of unions, intersections, and products of sets. Before going on, we will briefly review another way of axiomatizing set theory, which is of increasing interest in our day; the essential ideas are due to von Neumann.

Von Neumann noted that two facts combine to produce the logical paradoxes: in the first place, as we have seen, the crucial sets (for example, the set S in Russell's paradox) are "too large;" in the second place, these "large" sets *are allowed to be elements of sets* (for example, it is admitted that Russell's set S may be an element of itself). Of these two facts, Zermelo used the first; he avoided the paradoxes by making it impossible to form the "large" sets. Von

Neumann proposed to use the second of these facts: he would permit the excessively large sets to exist, but would not allow them to be elements of sets.

Briefly, von Neumann's system can be described as follows. As in Zermelo's theory, there is only one elementary predicate, namely the predicate $x \in Y$. The variables x, y, X, Y, etc. stand for objects which we agree to call *classes*; however, we distinguish between two kinds of classes, namely *elements*—which are defined to be those classes which are elements of classes—and *classes*, which are not elements of any class. Zermelo's axiom of selection is now replaced by a principle called the *class axiom*, which states the following:

> If $S(x)$ is any statement about an object x, there exists a class which consists of all those *elements* x which satisfy $S(x)$.

In other words, if $S(x)$ is any statement about x, we can form the class

$$\{x \mid x \text{ is an element and } S(x)\}.$$

To verify that Russell's paradox does not "work" in this system, let us go through the steps of Russell's argument, with S denoting $\{x \mid x \text{ is an element and } x \notin x\}$.

> $S \in S$ is impossible, for $S \in S$ implies $S \notin S$, which is a contradiction. Thus $S \notin S$. It follows that S is not an element, for, if S were an element, then we would have $S \in S$, which would be a contradiction.

Thus Russell's argument merely proves that the class S, defined above, is not an element.

The semantic paradoxes are avoided, as in the revised Zermelo system, by admitting in the class axiom only those "statements" which can be written in the formal language.

Variants of von Neumann's system have been developed by Gödel and Bernays. They have an advantage over Zermelo's system in that the class axiom is closer to the spirit of intuitive set theory than the axiom of selection. Indeed, if $S(x)$ is any statement about x, the class axiom guarantees the existence of a *class* containing all the elements x which satisfy $S(x)$. In mathematics, systems of the von Neumann type provide us with the convenience of being able to speak of the "class of all elements" and of being able to operate on classes which are not elements. (Such classes tend to occur at various points in higher mathematics; they can be avoided, but only at a price.) The chief disadvantage of these systems is that the distinction between classes which are elements and classes which are not elements—a highly artificial one—must always be borne in mind; however, this disadvantage is undoubtedly outweighed by the greater flexibility and naturalness of von Neumann type systems. In this text we shall use a slightly modified form of von Neumann's system of axioms.

5 OBJECTIONS TO THE AXIOMATIC APPROACH. OTHER PROPOSALS

What are the chief goals of axiomatic set theory, and to what extent have these goals been successfully attained? In order to answer that question we must remember the circumstances which led mathematicians, in the early years of the twentieth century, to search for an axiomatic basis to set theory. The ideas of Cantor had already thoroughly permeated the fabric of modern mathematics, and had become indispensable tools of the working mathematician. Algebra and analysis were formulated within a framework of set theory, and some of the most elegant, powerful new results in these fields were established by using the methods introduced by Cantor and his followers. Thus, when the paradoxes were discovered and there arose doubts as to the basic validity of Cantor's system, most mathematicians were understandably reluctant to give it up; they trusted that some way would be found to circumvent the contradictions and preserve, if not all, at least most of Cantor's results. Hilbert once wrote in this connection: "We will not be expelled from the paradise into which Cantor has led us."

With the discovery of newer paradoxes, and the failure of all the initial attempts to avoid them, it became increasingly clear that it would not be possible to preserve intuitive set theory in its entirety. Something—possibly quite a lot—would have to be relinquished. The best that one could hope for was to retain as much of intuitive set theory as was needed to save the new results of modern mathematics and provide an adequate framework for classical mathematics.

Briefly, then, axiomatic set theory was created to achieve a limited aim: it had to provide a firm foundation for a system of set theory which—while it did not need to be as comprehensive as intuitive set theory—must include all of Cantor's basic results as well as the constructions (such as the number systems, functions, and relations) needed for classical mathematics.

The systems of both Zermelo and von Neumann were successful in achieving this limited aim. But the amount of intuitive set theory which they had to sacrifice was considerable. For example, in Zermelo's system, as we have already seen, the intuitive way of making sets—by naming a property of objects and forming the set of all objects which have that property—does not take place at all. It is replaced by the axiom of selection, in which properties are allowed only to determine *subsets of given sets.* Furthermore, the only admissible "properties" are those which can be expressed entirely in terms of the seven symbols $\in$, $\vee$, $\wedge$, $\neg$, $\Rightarrow$, $\forall$, $\exists$ and variables x, y, z, ... As a result, many of the things that we normally think of as sets—for example, the "set of all apples," the "set of all atoms in the universe"—are not admissible as "sets" in axiomatic set theory. In fact, the only "sets" which the axioms provide for are, first, the empty set $\varnothing$, and then constructions such as $\{\varnothing\}$, $\{\varnothing, \{\varnothing\}\}$, etc., which can be built up from the empty set. It is a remarkable fact that all of mathematics can be based upon such a meager concept of set.

While the various axiomatic systems of set theory saved mathematics from its immediate peril, they failed to satisfy a great many people. In particular, many who were sensitive to the elegance and universality of mathematics were quick to point out that the creations of Zermelo and von Neumann must be regarded as provisional solutions—as expedients to solve a temporary problem; they will have to be replaced, sooner or later, by a mathematical theory of broader scope, which treats the concept of "set" in its full, intuitive generality.

This argument against axiomatic set theory—that it deals with an amputated version of our intuitive conception of a set—has important philosophical ramifications; it is part of a far wider debate, on the nature of mathematical "truth." The debate centers around the following question: Are mathematical concepts creations (that is, *inventions*) of the human mind, or do they exist independently of us in a "platonic" realm of concepts, merely to be *discovered* by the mathematician? The latter opinion is often referred to as "platonic realism" and is the dominant viewpoint of classical mathematics. We illustrate these two opposing points of view by showing how they apply to a particular concept—the notion of natural numbers. From the viewpoint of platonic realism, the concepts "one," "two," "three," and so on, exist in nature and existed before the first man began to count. If intelligent beings exist elsewhere in the universe, then, no matter how different they are from us, they have no doubt discovered the natural numbers and found them to have the same properties they have for us. On the other hand, according to the opposing point of view, while three cows, three stones, or three trees exist in nature, the natural number *three* is a creation of our minds; we have invented a procedure for constructing the natural numbers (by starting from zero and adding 1 each time, thus producing successively 1, 2, 3, etc.) and have in this manner fashioned a conceptual instrument of our own making.

How does platonic realism affect the status of axiomatic set theory? From the point of view of platonic realism, mathematical objects are given to us ready-made, with all their features and all their properties. It follows that to say a mathematical theorem is *true* means it expresses a correct statement about the relevant mathematical objects. (For example, the proposition $2 + 2 = 4$ is not merely a formal statement provable in arithmetic; it states an *actual fact* about numbers.) Now—if we admit that mathematical objects are given to us with all their properties, it follows, in particular, that the notion of *set* is a fixed, well-defined concept which we are not free to alter for our own convenience. Thus the "sets" created by Zermelo and von Neumann do not exist, and theorems which purport to describe these nonexistent objects are false! In conclusion, if we were to accept a strict interpretation of platonic realism, we would be forced to reject the systems of Zermelo and von Neumann as mathematically invalid.

Fortunately, the trend, for some time now, has been away from platonism and toward a more flexible, more "agnostic" attitude toward mathematical "truth." For one thing, developments in mathematics have been conforming less and less to the pattern dictated by platonic philosophy. For another, the cardinal requirement of platonism—that every mathematical object correspond to a definite, distinct object of our intuition (just as "point" and "line" refer to well-defined objects of our spatial intuition)—came to be an almost unbearable burden on the work of creative mathematicians by the nineteenth century. They were dealing with a host of new concepts (such as complex numbers, abstract laws of composition, and the general notion of function) which did not lend themselves to a simple interpretation in concrete terms. The case of the complex numbers is a good illustration of what was happening. Classical mathematics never felt at ease with the complex numbers, for it lacked a suitable "interpretation" of them, and as a result there were nagging doubts as to whether such things really "existed." Real numbers may be interpreted as lengths or quantities, but the square root of a negative real number—this did not seem to correspond to anything in the real world or in our intuition of number. Yet the system of the complex numbers arises in a most natural way—as the smallest number system which contains the real numbers and includes the roots of every algebraic equation with real coefficients; whether or not the complex numbers have a physical or psychological counterpart seems irrelevant.

The case of the complex numbers strikes a parallel with the problem of axiomatic set theory. For the "sets" created by Zermelo and von Neumann arise quite naturally in a mathematical context. They give us the simplest notion of set which is adequate for mathematics and yields a consistent axiomatic theory. Whether or not we can interpret them intuitively may be relatively unimportant.

Be that as it may, many mathematicians in the early 1900's were reluctant to make so sharp a break with tradition as axiomatic set theory seemed to demand. Furthermore, they felt, on esthetic grounds, that a mathematical theory of sets should describe all the things—and only those things—which our intuition recognizes to be sets. Among them was Bertrand Russell; in his efforts to reinstate intuitive set theory, Russell was led to the idea that we may consider sets to be ordered in a hierarchy of "levels," where, if A and $\mathscr{B}$ are sets and A is an element of $\mathscr{B}$, then $\mathscr{B}$ is "one level higher" than A. For example, in plane geometry, a *circle* (regarded as a set of points) is one level below a *family of circles*, which, in turn, is one level below a *set of families of circles*. This basic idea was built by Russell into a theory called the *theory of types*, which can be described, in essence, as follows.

Every set has a natural number assigned to it, called its *level*. The simplest sets, those of level 0, are called *individuals*—they do not have elements. A collection of individuals is a set of level 1; a collection of sets of level 1 is a set of level 2; and so on. In the theory of types the expression $a \in B$ is only meaningful if, for some number n, a is a set of level n and B is a set of level $n + 1$. It follows that the statement $x \in x$ has no meaning in the theory of types, and as a result, Russell's paradox vanishes for the simple reason that it cannot even be formulated.

Russell's theory of types is built upon a beautifully simple idea. Unfortunately, in order to make it "work," Russell was forced to add a host of new assumptions, until finally the resulting theory became too cumbersome to work with and too complicated to be truly pleasing. For one thing, corresponding to the hierarchy of "levels" of sets, it was necessary to have a hierarchy of "levels" of logical predicates. Then—as a way of avoiding the semantic paradoxes—sets at the same level were further divided into "orders." Finally, Russell had to admit a so-called *Axiom of Reducibility* which was just as arbitrary, just as ungrounded in intuition, as any of the *ad hoc* assumptions made by Zermelo. As a result of these shortcomings, the theory of types has not gained wide acceptance among mathematicians, although it is still an interesting (and maybe promising) area of research.

A far more radical approach was taken by a group of mathematicians calling themselves *intuitionists*. To the intuitionists, much of modern mathematics, including almost all of Cantor's theory of sets, is based on the uncritical use of rules of logic which they consider to be invalid. The intuitionist attitude toward set theory can therefore be summed up very easily: it is one of almost total rejection.

In order to properly understand the philosophy of intuitionism, we must first gain an understanding of its attitude toward logic. As the intuitionist sees it, the rules of logic used by mathematicians have an empirical character. Certain methods of proof came to be commonly used by mathematicians, and, over the years, were codified into a body of rules. These rules were observably correct in their original context, but—after they were codified—they came to be used uncritically in totally different contexts in which they no longer applied. Let us be more specific: in Euclidean geometry, which is the source of most mathematics before the fifteenth century, every theorem involves only a finite number of objects, and each of these objects (geometric figures) is given by an explicit construction. The rules of logic used by Euclid are perfectly valid in this context, say the intuitionists; it is only when they are transposed to problems involving an infinite domain of objects, or in which the objects are not given by an explicit construction, that the rules are incorrect.

As an example, let us take the *law of the excluded middle*. This is the rule which says that if S is any statement, then either S is true or the denial of S is true. In particular, let A be a set and let $P(x)$ be a statement which is meaningful for every element x in A. By the law of the excluded middle, either *there exists an* x *in* A *such that* $P(x)$ *is true*, or else *for every* x *in* A $P(x)$ *is false*. Now the intuitionists will accept this rule if A is a finite set and if each of its elements can be tested to determine whether $P(x)$ is true or false. In fact, say the intuitionists, this is where the rule originated—our experience tells us that if we examine every element x in A to determine whether or not $P(x)$ is true (to do so, A has to be a finite set), there can be only two possible outcomes: either we have found an x for which $P(x)$ is true, or else, for every x in A we have found $P(x)$ to be false. Our experience, then, confirms the law of the excluded middle in the case of finite sets. But, say the intuitionists, to assume it is true in the case of infinite sets—in an area where we have no experience and experience is impossible—is wholly without foundation. On grounds such as these, the intuitionists deny the law of the excluded middle. Other rules of traditional logic are similarly rejected, because they go beyond the realm of our experience.

The intuitionist points out that mathematics originated as a study of certain mental constructions—chiefly geometric figures and simple constructions involving whole numbers. The theorems of early mathematics were essentially statements to the effect that if certain constructions are carried out, certain results will be achieved. For example, consider the following theorem of geometry: Given two triangles, if two sides and the included angle of one triangle are equal, respectively, to two sides and the included angle of the other triangle, then the two triangles are congruent. What is expressed here is the fact that if we construct two triangles, with two sides and an included angle equal as stated above, we will be able to verify (for example, by using a compass) that the remaining side of one triangle is equal to the remaining side of the other triangle. The proofs of Euclidean geometry, and of early mathematics generally, have a constructive character. For example, the Pythagorean theorem is proven by constructing a figure in which corresponding parts are congruent, hence have the same area. Once the construction is completed, it remains only to point out the congruent parts and thereby reach the desired conclusion. Thus, say the intuitionists, rules of logic were originally intended to describe situations which arose in the context of such constructions and determinations. For example, the rule of the excluded middle was intended simply to note the fact that if we are given a (finite) set of objects and a method (for example, using a ruler and compass) to test each object for some property P, then, if we perform the test on every object, either one of the objects will pass the requirements of the test, or else every object will fail the requirements.

To sum up, then, the intuitionist maintains that a mathematical theorem is nothing more than a factual statement to the effect that a given mental construction will lead to a given result. Every proof must be constructive. If we claim that a mathematical object exists, we must prove it by giving a method for actually constructing the object. If we assert that a relation holds among given pairs of objects, our proof must include a method for testing every pair of objects in question. The "rules of logic" are nothing more than simple observations on the process of performing mathematical constructions; we have no grounds to believe these rules apply outside the context of constructive mathematics—in fact, it is meaningless to apply them outside this context. *Logic is incidental, not essential, to mathematics.*

It is obvious that the intuitionist's notion of *set* must be quite different from ours. Consider, for example, Cantor's principle that if we can name a property of objects, then there exists a set of all objects which have that property. Now this principle—as well as the limited versions of it accepted by Zermelo and von Neumann—is anathema to the intuitionists. An object exists only if it can be constructed; hence, a set exists only if we are able to describe a procedure for building it.

A full discussion of intuitionist set theory is outside the scope of this book; however, it is worth mentioning a particular kind of set which is important in intuitionist mathematics: this is called a *spread.* A spread is identified with a rule for producing all of its elements. Thus, a spread is not regarded as an "already formed" totality, but rather as a "process of formation"; each of its elements can be formed if we apply the rule long enough.

The paradoxes do not occur in intuitionist set theory because the crucial sets in the paradoxes cannot be produced in intuitionist mathematics, and the essential arguments cannot be rendered using intuitionist logic.

6 CONCLUDING REMARKS

During the early part of the twentieth century, as we have seen, various ways of building a noncontradictory theory of sets were proposed and developed by different "schools" of mathematicians. We have reviewed the basic principles of axiomatic set theory, Russell's theory of types and the intuitionist (or "constructivist") approach to sets. In addition to these, a great many other ideas were proposed, too various to describe in this brief introduction.

Of all the ways of dealing with sets, the axiomatic method seemed to best suit the needs of modern mathematics. The notion of "set" embodied in the systems of Zermelo and von Neumann is broad enough for the purposes of mathematics, and therefore in a mathematical setting it is virtually indistinguishable from the Cantorian notion of set. The methods of proof, the symbolism, the rigor—all of these correspond to current mathematical usage. Most important of all, axiomatic set theory seems "natural" to most working mathematicians.

Those who reject axiomatic set theory do so on the basis of some philosophical bias. Those philosophical positions, however, which refuse to accept axiomatic set theory also deny the validity of a large part of modern mathematics. For example, the intuitionist school rejects the greater part of contemporary analysis, because it is founded on nonconstructivist principles. While the arguments of these critics present a challenge, and certainly give us food for thought, they are not powerful enough to destroy the achievements of three brilliant generations of mathematicians.

In the course of the past seventy years or so, set theory has come to be widely recognized as the fundamental, "unifying" branch of mathematics. We have already seen how the natural numbers can be constructed, and their properties derived, within the framework of set theory; from there, it is easy to develop the rational numbers, the real and complex numbers, as well as remarkable systems such as Cantor's "transfinite cardinals." The notions of function, relation, operation, and so forth are easily defined in terms of sets, and, as a result, every known branch of mathematics can be formulated within set theory. It is therefore legitimate—and, in fact, vital—to ask the question: "How secure a foundation does set theory provide for the whole edifice of mathematics?" In particular, are we absolutely certain that axiomatic set theory is consistent, that is, free of contradictions? If it is, then everything we develop within it—in other words, all of mathematics—is consistent; and if it is not, then whatever we build upon it is worthless.

The fact is that there is no known proof of the consistency of axiomatic set theory. This is not too surprising though, in view of some of the results of modern logic. For example, in 1931, K. Gödel proved that it is impossible to give a finitary proof of the consistency of ordinary arithmetic (of the natural numbers). The situation in almost every other branch of mathematics is much the same. Thus the best assurance that we have, at this time, of the consistency of axiomatic set theory is the fact that the familiar contradictions cannot be obtained in the usual way. We cannot do any better at this time.

A result relating to *relative consistency* is of some interest. It has been proven recently that *if* Zermelo's axiomatization of set theory is consistent, *then* von Neumann's axiomatization is also consistent.

It is probable that, in the final analysis, any assurance of the consistency of mathematics will have to rest on some combination of basic intuition and empirical evidence.

1
Classes and Sets

1 BUILDING SENTENCES

Before introducing the basic notions of set theory, it will be useful to make certain observations on the use of language.

By a *sentence* we will mean a statement which, in a given context, is unambiguously either *true* or *false*. Thus

London is the capital of England.

Money grows on trees.

Snow is black.

are examples of sentences. We will use letters P, Q, R, S, etc., to denote sentences; used in this sense, P, for instance, is to be understood as asserting that "P is true."

Sentences may be combined in various ways to form more complicated sentences. Often, the truth or falsity of the compound sentence is completely determined by the truth or falsity of its component parts. Thus, if P is a sentence, one of the simplest sentences we may form from P is the **negation** of P, denoted by $\neg P$ (to be read "not P"), which is understood to assert that "P *is false.*" Now if P is true, then, quite clearly, $\neg P$ is false; and if P is false, then $\neg P$ is true. It is convenient to display the relationship between $\neg P$ and P in the following *truth table*,

1.1

P	$\neg P$
t	f
f	t

where t and f denote the "truth values", *true* and *false*.

Another simple operation on sentences is conjunction: if P and Q are sentences, the **conjunction** of P and Q, denoted by $P \wedge Q$ (to be read "P and Q"), is understood to assert that "P *is true and* Q *is true.*" It is intuitively clear that $P \wedge Q$ is true if P and Q are both true, and false otherwise; thus, we have the following truth table.

1.2

P	Q	$P \wedge Q$
t	t	t
t	f	f
f	t	f
f	f	f

The **disjunction** of P and Q, denoted by $P \vee Q$ (to be read "P or Q"), is the sentence which asserts that "*P is true, or Q is true, or P and Q are both true.*" It is clear that $P \vee Q$ is false only if P and Q are both false.

1.3

P	Q	$P \vee Q$
t	t	t
t	f	t
f	t	t
f	f	f

An especially important operation on sentences is **implication**: if P and Q are sentences, then $P \Rightarrow Q$ (to be read "P implies Q") asserts that "*if P is true, then Q is true.*" *A word of caution*: in ordinary usage, "if P is true, then Q is true" is understood to mean that there is a *causal* relationship between P and Q (as in "if John passes the course, then John can graduate"). In mathematics, however, implication is always understood in the *formal* sense: $P \Rightarrow Q$ is true *except if* P is true and Q is false. In other words, $P \Rightarrow Q$ is defined by the truth table.

1.4

P	Q	$P \Rightarrow Q$
t	t	t
t	f	f
f	t	t
f	f	t

The properties of formal implication differ somewhat from the properties we would expect "causal" implication to have. For example,

$$\text{“}1 + 1 = 2 \quad \Rightarrow \quad \pi \text{ is a transcendental number”}$$

is true, even though there is no causal relationship between the two component sentences. To take another example,

$$\text{“}2 + 2 = 5 \quad \Rightarrow \quad 4 \text{ is a prime number”}$$

is true, even though the two component sentences are false. This should not disturb the reader unduly, for formal implication still has the fundamental property which we demand of implication—namely, if $P \Rightarrow Q$ is true, then, necessarily, if P is true then Q is true.

Certain compound sentences are true regardless of the truth or falsity of their component parts; a typical example is the sentence $P \Rightarrow P$. Regardless of whether P is true or false, $P \Rightarrow P$ is always true; in other words, no matter what sentence P is, $P \Rightarrow P$ is true. For future reference we record a few sentences which have this property.

1.5 Theorem For all sentences P and Q, the following statements are true.

i) $P \Rightarrow P \vee Q$. i)′ $Q \Rightarrow P \vee Q$.
ii) $P \wedge Q \Rightarrow P$. ii)′ $P \wedge Q \Rightarrow Q$.

Proof

i) We wish to prove that if P and Q are any sentences, then $P \Rightarrow P \vee Q$ is true; in other words, we wish to prove that no matter what truth values are assumed by P and Q, $P \Rightarrow P \vee Q$ is always true. To do this, we derive a truth table for $P \Rightarrow P \vee Q$ as follows.

P	Q	$P \vee Q$	$P \Rightarrow P \vee Q$
t	t	t	t
t	f	t	t
f	t	t	t
f	f	f	t

The basic idea of the derived truth table is this: in line 1, P and Q both take the value t; thus, by 1.3, $P \vee Q$ takes the value t; now, P has the value t and $P \vee Q$ has the value t, so, by 1.4, $P \Rightarrow P \vee Q$ takes the value t. We do the same for each line, and we find that in every line (that is, for every possible assignment of truth values to P and Q) $P \Rightarrow P \vee Q$ has the value t (true). This is what we had set out to prove.

i)′ The derived truth table for $Q \Rightarrow P \vee Q$ is analogous to the one for $P \Rightarrow P \vee Q$; the conclusion is the same.

ii) In order to prove that $P \wedge Q \Rightarrow P$ for all sentences P and Q, we derive a truth table for $P \wedge Q \Rightarrow P$.

P	Q	$P \wedge Q$	$P \wedge Q \Rightarrow P$
t	t	t	t
t	f	f	t
f	t	f	t
f	f	f	t

In every line (that is, for every possible assignment of truth values to P and Q), $P \wedge Q \Rightarrow P$ takes the value t; thus, $P \wedge Q \Rightarrow P$ is true irrespective of the truth or falsity of its component sentences P and Q.

ii)′ The truth table for $P \wedge Q \Rightarrow Q$ is analogous to the one for $P \wedge Q \Rightarrow P$, and the conclusion is the same. ■

1.6 Theorem For all sentences P, Q and R, the following is true:

$$[(P \Rightarrow Q) \wedge (Q \Rightarrow R)] \Rightarrow (P \Rightarrow R).$$

Proof The reader should derive the truth table for

$$[(P \Rightarrow Q) \wedge (Q \Rightarrow R)] \quad \Rightarrow \quad (P \Rightarrow R)$$

and verify that this sentence takes the truth value t in every line of the table. ■

1.7 Theorem For all sentences P, Q and R, if $Q \Rightarrow R$ is true, then

i) $P \vee Q \Rightarrow P \vee R$ is true, and

ii) $P \wedge Q \Rightarrow P \wedge R$ is true.

Proof

i) We assume that $Q \Rightarrow R$ is true, and derive the truth table for $P \vee Q \Rightarrow P \vee R$.

P	Q	R	$P \vee Q$	$P \vee R$	$P \vee Q \Rightarrow P \vee R$
t	t	t	t	t	t
t	t	f	t	t	t
t	f	t	t	t	t
t	f	f	t	t	t
f	t	t	t	t	t
f	t	f	t	f	f
f	f	t	f	t	t
f	f	f	f	f	t

Since we assume that $Q \Rightarrow R$ is true, we cannot have, simultaneously, Q true and R false; thus, we may disregard the sixth line of the table. In all of the remaining lines, $P \vee Q \Rightarrow P \vee R$ takes the value t.

ii) The proof that $P \wedge Q \Rightarrow P \wedge R$ is analogous to the above. ■

We agree that $P \Leftrightarrow Q$ is to be an abbreviation for $(P \Rightarrow Q) \wedge (Q \Rightarrow P)$.

1.8 Theorem For all sentences P, Q and R, the following are true:

i) $P \vee Q \Leftrightarrow Q \vee P$,

ii) $P \vee (Q \vee R) \Leftrightarrow (P \vee Q) \vee R$,

iii) $P \wedge (Q \vee R) \Leftrightarrow (P \wedge Q) \vee (P \wedge R)$,

iv) $P \vee P \Leftrightarrow P$,

i)′ $P \wedge Q \Leftrightarrow Q \wedge P$,

ii)′ $P \wedge (Q \wedge R) \Leftrightarrow (P \wedge Q) \wedge R$,

iii)′ $P \vee (Q \wedge R) \Leftrightarrow (P \vee Q) \wedge (P \vee R)$,

iv)′ $P \wedge P \Leftrightarrow P$.

The proof of this theorem is left as an exercise for the reader.

In this and the subsequent chapters, $\Rightarrow$ will be used as an abbreviation for *implies*, $\Leftrightarrow$ will be used as an abbreviation for *if and only if* (we will sometimes write "iff" instead of $\Leftrightarrow$), $\wedge$ will be used as an abbreviation for *and*, and $\vee$ will be used as an abbreviation for *or*. If $P, Q, R, \ldots$ are any statements, an expression of the form $P \Rightarrow Q \Rightarrow R \Rightarrow \cdots$ should be understood to mean that $P \Rightarrow Q$, $Q \Rightarrow R$, and so on; analogously, $P \Leftrightarrow Q \Leftrightarrow R \Leftrightarrow \cdots$ should be understood to mean that $P \Leftrightarrow Q$, $Q \Leftrightarrow R$, and so on.

As is customary, $\exists$ is to be read *there exists*, $\forall$ is to be read *for all*, and $\ni$ is to be read *such that*.

EXERCISES 1.1

1. Prove Theorem 1.8.
2. Prove that the following sentences are true for all P and Q (DeMorgan's Laws).

   a) $\neg(P \vee Q) \Leftrightarrow \neg P \wedge \neg Q$. b) $\neg(P \wedge Q) \Leftrightarrow \neg P \vee \neg Q$.
3. Prove that the following sentences are true, for every sentence P.

   a) $\neg\neg P \Rightarrow P$. b) $P \Rightarrow \neg\neg P$.
4. Prove that the following sentences are true for all P and Q.

   a) $(P \Rightarrow Q) \Leftrightarrow (\neg Q \Rightarrow \neg P)$. b) $(P \Rightarrow Q) \Leftrightarrow (\neg P \vee Q)$.
   c) $(P \Rightarrow Q) \Leftrightarrow \neg(P \wedge \neg Q)$. d) $[P \wedge (P \Rightarrow Q)] \Rightarrow Q$.
   e) $[(P \vee Q) \wedge \neg P] \Rightarrow Q$.
5. Prove the following sentences are true for all P, Q and R.

   a) $[(P \Rightarrow Q) \wedge (Q \Rightarrow R)] \Rightarrow (P \Rightarrow R)$.
   b) $[(P \Rightarrow Q) \wedge (R \Rightarrow Q)] \Leftrightarrow [(P \vee R) \Rightarrow Q]$.
   c) $[(P \Rightarrow Q) \wedge (P \Rightarrow R)] \Leftrightarrow [P \Rightarrow (Q \wedge R)]$.
6. Prove that, for all sentences P, Q and R, if $Q \Leftrightarrow R$ is true, then the following are true.

   a) $P \vee Q \Leftrightarrow P \vee R$. b) $P \wedge Q \Leftrightarrow P \wedge R$.
   c) $(P \Rightarrow Q) \Leftrightarrow (P \Rightarrow R)$.
7. Prove that for all sentences P, Q, R and S, if $P \Rightarrow Q$ and $R \Rightarrow S$, then

   a) $P \vee R \Rightarrow Q \vee S$, b) $P \wedge R \Rightarrow Q \wedge S$.

2 BUILDING CLASSES

We will now begin our development of axiomatic set theory.

Every axiomatic system, as we have seen, must start with a certain number of *undefined notions*. For example, in geometry, the words "point" and "line" and the relation of "incidence" are generally taken to be undefined. While we are free in our own minds to attach a "meaning," in the form of a mental picture, to each of these notions, mathematically we must proceed "as if" we did not know what they meant. Now an "undefined" notion has no properties except those which are explicitly assigned to it; therefore, we must state as *axioms* all the elementary properties which we expect our undefined notions to have.

In our system of axiomatic set theory we choose two undefined notions: the word *class* and the *membership relation* $\in$. All the objects of our theory are called classes. Certain classes, to be called *sets*, will be defined in Section 7. Every set is a class, but not conversely; a class which is not a set is called a *proper class*.

Let us comment briefly on the "meaning" we intend to attach to these notions. In the intended interpretation of our axiomatic system, the word *class* is understood to refer to any collection of objects. However, as we noted in section 2 of Chapter 0, certain "excessively large" collections can be formed in intuitive set theory (for example, the collection of all x such that $x \notin x$), and if we do not exercise special caution they lead to contradictions such as Russell's paradox. The term *proper class* is understood to refer to these "excessively large" collections; all other collections are *sets*.

If x and A are classes, the expression $x \in A$ is read "x is an element of A," or "x belongs to A," or "x is in A." It is convenient to write $x \notin A$ for "x is not an element of A." Let x be a class; if there exists a class A such that $x \in A$, then x is called an *element*.

From here on we will use the following notational convention: **lower-case letters** $a, b, c, x, y, \ldots$ **will be used only to designate elements.** Thus, a capital letter, such as A, may denote either an element or a class which is not an element, but a lower-case letter, such as x, may denote *only* an element.

1.9 Definition Let A and B be classes; we define $A = B$ to mean that every element of A is an element of B and vice versa. In symbols,

$$A = B \quad \text{iff} \quad x \in A \Rightarrow x \in B \;\text{ and }\; x \in B \Rightarrow x \in A.$$

We have defined two classes to be equal if and only if they have the same elements. Equal classes have another property: if x and y are equal and x is an element of A, we certainly expect y to be an element of A. This property is stated as our first axiom:

A1. If $x = y$ and $x \in A$, then $y \in A$.

This axiom is sometimes called the *axiom of extent.*

1.10 Definition Let A and B be classes; we define $A \subseteq B$ to mean that every element of A is an element of B. In symbols,

$$A \subseteq B \quad \text{iff} \quad x \in A \Rightarrow x \in B.$$

If $A \subseteq B$, then we say that A is a *subclass* of B.

We define $A \subset B$ to mean that $A \subseteq B$ and $A \neq B$; in this case, we say that A is a *strict subclass* of B.

If A is a subclass of B, and A is a set, we will call A a *subset* of B.

A few simple properties of equality and inclusion are given in the next theorem.

1.11 Theorem For all classes A, B and C, the following hold:

i) $A = A$.
ii) $A = B \Rightarrow B = A$.
iii) $A = B$ and $B = C \Rightarrow A = C$.
iv) $A \subseteq B$ and $B \subseteq A \Rightarrow A = B$.
v) $A \subseteq B$ and $B \subseteq C \Rightarrow A \subseteq C$.

Proof

i) The statement $x \in A \Rightarrow x \in A$ and $x \in A \Rightarrow x \in A$ is obviously true; thus, by Definition 1.9, $A = A$.

ii) Suppose $A = B$; then $x \in A \Rightarrow x \in B$ and $x \in B \Rightarrow x \in A$; hence by 1.8(i)′ $x \in B \Rightarrow x \in A$ and $x \in A \Rightarrow x \in B$; thus, by Definition 1.9, $B = A$.

iii) Suppose $A = B$ and $B = C$; then we have the following:

$$x \in A \Rightarrow x \in B,$$
$$x \in B \Rightarrow x \in A,$$
$$x \in B \Rightarrow x \in C,$$
$$x \in C \Rightarrow x \in B.$$

From the first and third of these statements we conclude (by 1.6) that $x \in A \Rightarrow x \in C$. From the second and fourth of these statements we conclude that $x \in C \Rightarrow x \in A$. Thus, by Definition 1.9, $A = C$.

We leave the proofs of (iv) and (v) as an exercise for the reader. ■

We have seen that the intuitive way of making classes is to name a property of objects and form the class of all the objects which have that property. Our second axiom allows us to make classes in this manner.

A2. Let $P(x)$ designate a statement about x which can be expressed entirely in terms of the symbols $\in$, $\vee$, $\wedge$, $\neg$, $\Rightarrow$, $\exists$, $\forall$, brackets, and variables x, y, z, A, B, ... Then there exists a class C which consists of all the elements x which satisfy $P(x)$.

Axiom A2 is called the *axiom of class construction.*

The reader should note that axiom A2 permits us to form the class of all the *elements* x which satisfy $P(x)$, *not* the class of all the *classes* x which satisfy $P(x)$; as we discussed on page 13, this distinction is sufficient to eliminate the logical paradoxes. The semantic paradoxes have been avoided by ad. .itting in axiom A2 only those statements $P(x)$ which can be written entirely in terms of the symbols $\in$, $\vee$, $\wedge$, $\neg$, $\Rightarrow$, $\exists$, $\forall$, brackets and variables.

The class C whose existence is asserted by Axiom A2 will be designated by the symbol

$$C = \{x \mid P(x)\}.$$

1.12 *Remark.* The use of a small x in the expression $\{x \mid P(x)\}$ is not accidental, but quite essential. Indeed, we have agreed that lower-case letters x, y, etc., will be used *only* to designate elements. Thus

$$C = \{x \mid P(x)\}$$

asserts that C is the class of all the *elements* x which satisfy $P(x)$.

We will now use the axiom of class construction to build some new classes from given classes.

1.13 Definition Let A and B be classes; the *union* of A and B is defined to be the class of all the elements which belong either to A, or to B, or to both A and B. In symbols,

$$A \cup B = \{x \mid x \in A \text{ or } x \in B\}.$$

Thus, $x \in A \cup B$ if and only if $x \in A$ or $x \in B$.

1.14 Definition Let A and B be classes; the *intersection* of A and B is defined to be the class of all the elements which belong to both A and B. In symbols,

$$A \cap B = \{x \mid x \in A \text{ and } x \in B\}.$$

Thus, $x \in A \cap B$ if and only if $x \in A$ and $x \in B$.

1.15 Definition By the *universal class* $\mathscr{U}$ we mean the class of all elements. The existence of the universal class is a consequence of the axiom of class construction, for if we take $P(x)$ to be the statement $x = x$, then A2 guarantees the existence of a class which consists of all the elements which satisfy $x = x$; by 1.11(i), every element is in this class.

1.16 Definition By the *empty class* we mean the class $\varnothing$ which has no elements at all. The existence of the empty class is a consequence of the axiom of class construction; indeed, A2 guarantees the existence of a class which consists of all the elements which satisfy $x \neq x$; by Theorem 1.11(i), this class has no elements.

1.17 Theorem For every class A, the following hold:

i) $\varnothing \subseteq A$. ii) $A \subseteq \mathscr{U}$.

Proof

i) In order to prove that $\varnothing \subseteq A$, we must show that $x \in \varnothing \Rightarrow x \in A$. It suffices to prove the contrapositive of this statement, that is, $x \notin A \Rightarrow x \notin \varnothing$. Well, suppose $x \notin A$; then certainly $x \notin \varnothing$, for $\varnothing$ has no elements; thus $x \notin A \Rightarrow x \notin \varnothing$.

ii) If $x \in A$, then x is an element; hence $x \in \mathscr{U}$. ■

1.18 Definition If two classes have no elements in common, they are said to be *disjoint*. In symbols,

$$A \text{ and } B \text{ are disjoint} \quad \text{iff} \quad A \cap B = \varnothing.$$

1.19 Definition The *complement* of a class A is the class of all the elements which do not belong to A. In symbols,

$$A' = \{x \mid x \notin A\}.$$

Thus, $x \in A'$ if and only if $x \notin A$.

Relations among classes can be represented graphically by means of a useful device known as the *Venn diagram.* A class is represented by a simple plane area (circular or oval in shape); if it is desired to show the complement of a class, then the circle or oval is drawn within a rectangle which represents the universal class. Thus, $A \cup B$ is rendered by the shaded area of Fig. 1, $A \cap B$ by the shaded area of Fig. 2, and A' by the shaded area of Fig. 3. The reader will find that Venn diagrams are helpful in guiding his reasoning about classes, and that they give more meaning to set-theoretic formulas by making them more concrete. For example, in Section 3 of this chapter we will prove the formula

$$A \cap (B \cup C) = (A \cap B) \cup (A \cap C).$$

This formula is illustrated in Fig. 4, where the shaded area represents

$A \cap (B \cup C)$; one immediately notices that this same shaded area represents $(A \cap B) \cup (A \cap C)$.

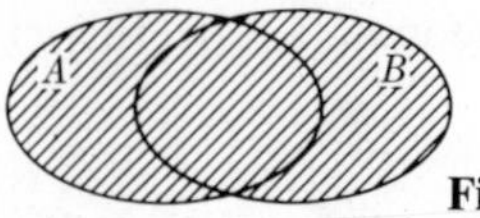

Fig. 1

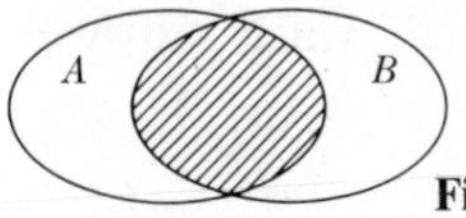

Fig. 2

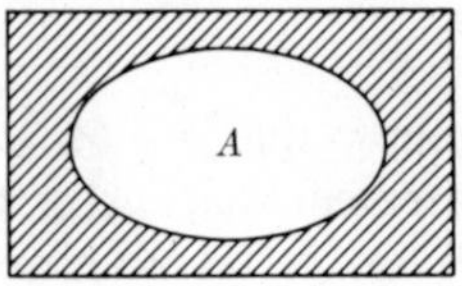

Fig. 3

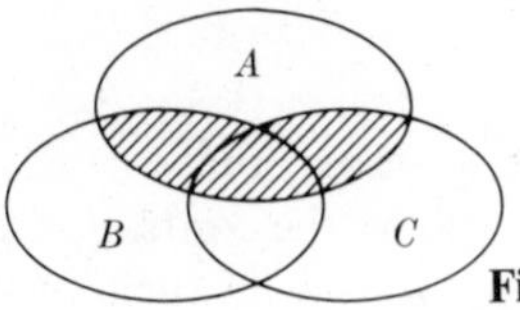

Fig. 4

EXERCISES 1.2

1. Suppose that $A \subseteq B$ and $C \subseteq D$; prove that

   a) $(A \cup C) \subseteq (B \cup D)$, b) $(A \cap C) \subseteq (B \cap D)$.

   [*Hint:* Use the result of Exercise 7, Exercise Set 1.1.]

2. Suppose $A = B$ and $C = D$; prove that

   a) $A \cup C = B \cup D$, b) $A \cap C = B \cap D$.

   [*Hint:* Use the result of the preceding exercise.]

3. Prove that if $A \subseteq B$, then $B' \subseteq A'$.

   [*Hint:* Use the result of Exercise 4(a), Exercise Set 1.1.]

4. Prove that if $A = B$, then $A' = B'$.
5. Prove that if $A = B$ and $B \subseteq C$, then $A \subseteq C$.
6. Prove that if $A \subset B$ and $B \subset C$, then $A \subset C$.
7. Prove Theorem 1.11, parts (iv) and (v).
8. Let $S = \{x \mid x \notin x\}$; use Russell's argument to prove that S is not an element.
9. Does Axiom A2 allow us to form the "class of all classes"? Explain.
10. Explain why Russell's paradox and Berry's paradox cannot be produced by using Axiom A2.

3 THE ALGEBRA OF CLASSES

One of the most interesting and useful facts about classes is that under the operations of union, intersection, and complementation they satisfy certain algebraic laws from which we can develop an algebra of classes. We shall see later (Chapter 4) that the algebra of classes is merely one example of a structure known as a *Boolean algebra*; another example is the "algebra of logic," where $\vee$, $\wedge$, $\neg$ are regarded as operations on sentences.

Our purpose in this section is to develop the basic laws of the algebra of classes. We remind the reader that the word *class* should be understood to mean *any collection of objects*; thus, the laws we are about to present should be thought of as applying to every collection of objects; in particular, they apply to all sets.

1.20 Theorem If A and B are any classes, then

i) $A \subseteq A \cup B$ and $B \subseteq A \cup B$,

ii) $A \cap B \subseteq A$ and $A \cap B \subseteq B$.

Proof

i) To prove that $A \subseteq A \cup B$, we must show that $x \in A \Rightarrow x \in A \cup B$:

$$x \in A \Rightarrow x \in A \vee x \in B \quad \text{by 1.5(i)}$$

$$\Rightarrow x \in A \cup B \quad \text{by 1.13.}$$

Analogously, we can show that $B \subseteq A \cup B$.

ii) To prove that $A \cap B \subseteq A$, we must show that $x \in A \cap B \Rightarrow x \in A$.

$$x \in A \cap B \Rightarrow x \in A \wedge x \in B \quad \text{by 1.14}$$

$$\Rightarrow x \in A \blacksquare \quad \text{by 1.5(ii).}$$

1.21 Theorem If A and B are classes, then

i) $A \subseteq B$ if and only if $A \cup B = B$,

ii) $A \subseteq B$ if and only if $A \cap B = A$.

Proof

i) Let us first assume that $A \subseteq B$; that is, $x \in A \Rightarrow x \in B$. Then

$$x \in A \cup B \Rightarrow x \in A \quad \text{or} \quad x \in B \quad \text{by 1.13}$$

$$\Rightarrow x \in B \quad \text{or} \quad x \in B \quad \text{by 1.7(i)}$$

$$\Rightarrow x \in B \quad \text{by 1.8(iv).}$$

Thus, $A \cup B \subseteq B$; but $B \subseteq A \cup B$ by 1.20(i); consequently, $A \cup B = B$.

Conversely, let us assume that $A \cup B = B$. By 1.20(i), $A \subseteq A \cup B$; thus $A \subseteq B$.

ii) The proof is left as an exercise for the reader. ■

1.22 Theorem (*Absorption Laws*). For all classes A and B,

i) $A \cup (A \cap B) = A$. ii) $A \cap (A \cup B) = A$.

Proof

i) By 1.20(ii), $A \cap B \subseteq A$; therefore, by 1.21(i), $A \cup (A \cap B) = A$.

ii) By 1.20(i), $A \subseteq A \cup B$; therefore, by 1.21(ii), $A \cap (A \cup B) = A$. ■

1.23 Theorem For every class A, $(A')' = A$.

Proof

$$x \in (A')' \Rightarrow x \notin A' \Rightarrow x \in A \qquad \text{by 1.19,}$$
$$x \in A \Rightarrow x \notin A' \Rightarrow x \in (A')' \qquad \text{by 1.19.} ■$$

1.24 Theorem (*DeMorgan's Laws*). For all classes A and B,

i) $(A \cup B)' = A' \cap B'$. ii) $(A \cap B)' = A' \cup B'$.

Proof

i) First, $x \in (A \cup B)' \Rightarrow x \notin A \cup B$ — by 1.19

$\Rightarrow x \notin A$ and $x \notin B$

(because if either $x \in A$ or $x \in B$, then $x \in A \cup B$)

$\Rightarrow x \in A'$ and $x \in B'$ — by 1.19

$\Rightarrow x \in (A' \cap B')$ — by 1.14.

Next, $x \in (A' \cap B') \Rightarrow x \in A'$ and $x \in B'$ — by 1.14

$\Rightarrow x \notin A$ and $x \notin B$ — by 1.19

$\Rightarrow x \notin A \cup B$

$\Rightarrow x \in (A \cup B)'$ — by 1.19.

ii) The proof is left as an exercise for the reader. ■

1.25 Theorem For all classes A, B and C, the following are true.

Commutative Laws:

i) $A \cup B = B \cup A$

ii) $A \cap B = B \cap A$

Idempotent Laws:

iii) $A \cup A = A$

iv) $A \cap A = A$

Associative Laws:

v) $A \cup (B \cup C) = (A \cup B) \cup C$

vi) $A \cap (B \cap C) = (A \cap B) \cap C$

Distributive Laws:
vii) $A \cap (B \cup C) = (A \cap B) \cup (A \cap C)$
viii) $A \cup (B \cap C) = (A \cup B) \cap (A \cup C)$

Proof

i) $x \in A \cup B \Rightarrow x \in A \text{ or } x \in B$ by 1.13
$\Rightarrow x \in B \text{ or } x \in A$ by 1.8(i)
$\Rightarrow x \in B \cup A$ by 1.13.

v) $x \in A \cup (B \cup C) \Rightarrow x \in A \vee x \in B \cup C$ by 1.13
$\Rightarrow x \in A \vee (x \in B \vee x \in C)$ by 1.13
$\Rightarrow (x \in A \vee x \in B) \vee x \in C$ by 1.8(ii)
$\Rightarrow x \in A \cup B \vee x \in C$ by 1.13
$\Rightarrow x \in (A \cup B) \cup C$ by 1.13.

vii) $x \in A \cap (B \cup C) \Rightarrow x \in A \wedge x \in B \cup C$ by 1.14
$\Rightarrow x \in A \wedge (x \in B \vee x \in C)$ by 1.13
$\Rightarrow (x \in A \wedge x \in B) \vee (x \in A \wedge x \in C)$ by 1.8(iii)
$\Rightarrow x \in A \cap B \vee x \in A \cap C$ by 1.14
$\Rightarrow x \in (A \cap B) \cup (A \cap C)$ by 1.13.

The proofs of (ii), (iii), (iv), (vi), and (viii) are exercises for the reader. ■

The empty class and the universal class are identity elements for union and intersection respectively; they satisfy the following simple rules:

1.26 Theorem For every class A,

i) $A \cup \varnothing = A$. ii) $A \cap \varnothing = \varnothing$.
iii) $A \cup \mathcal{U} = \mathcal{U}$. iv) $A \cap \mathcal{U} = A$.
v) $\mathcal{U}' = \varnothing$. vi) $\varnothing' = \mathcal{U}$.
vii) $A \cup A' = \mathcal{U}$. viii) $A \cap A' = \varnothing$.

Proof

i) By 1.17, $\varnothing \subseteq A$, and therefore by 1.21(i), $A \cup \varnothing = A$.
iii) By 1.17(ii), $A \subseteq \mathcal{U}$, and therefore by 1.21(i), $A \cup \mathcal{U} = A$.

The proofs of the remaining parts of this theorem are left as an exercise for the reader. ■

By using the laws of class algebra which we have developed above, we can prove all the elementary properties of classes without referring to the definitions of the symbols $\cup$, $\cap$, $'$, and $\subseteq$. The following is an example of how such proofs are carried out.

Example Prove that $A \cap (A' \cup B) = A \cap B$.

Proof

$$\begin{aligned} A \cap (A' \cup B) &= (A \cap A') \cup (A \cap B) && \text{by 1.25(vii)} \\ &= \varnothing \cup (A \cap B) && \text{by 1.26(viii)} \\ &= A \cap B && \text{by 1.26(i).} \end{aligned}$$

The following definition is frequently useful: The *difference* of two classes A and B is the class of all elements which belong to A, but do not belong to B. In symbols,

$$A - B = A \cap B'.$$

Example Prove that $A - B = B' - A'$.

Proof

$$\begin{aligned} A - B &= A \cap B' && \text{Definition} \\ &= B' \cap A && \text{by 1.25(ii)} \\ &= B' \cap (A')' && \text{by 1.23} \\ &= B' - A' && \text{Definition of } B' - A'. \end{aligned}$$

It is useful to note that with the aid of Theorem 1.21, relations involving inclusion ($\subseteq$), not merely equality, can be proved using class algebra.

EXERCISES 1.3

1. Prove Theorem 1.21(ii).
2. Prove Theorem 1.24(ii).
3. Prove Theorem 1.25, parts (ii), (iii), (iv), (vi) and (viii).
4. Prove Theorem 1.26, parts (ii), (iv), (v) through (viii).
5. Use class algebra to prove the following.

   a) $(A \cap B) \cup C = (A \cup C) \cap (B \cup C)$, b) $(A \cup B) \cap C = (A \cap C) \cup (B \cap C)$.
6. Use class algebra to prove the following.

   a) If $A \cap C = \varnothing$, then $A \cap (B \cup C) = A \cap B$.
   b) If $A \cap B = \varnothing$, then $A - B = A$.
   c) If $A \cap B = \varnothing$ and $A \cup B = C$, then $A = C - B$.
7. Using class algebra, prove each of the following.

   a) $A \cap (B - C) = (A \cap B) - C$.
   b) $(A \cup B) - C = (A - C) \cup (B - C)$.

c) $A - (B \cup C) = (A - B) \cap (A - C)$.
d) $A - (B \cap C) = (A - B) \cup (A - C)$.

8. We define the operation $+$ on classes as follows: If A and B are classes, then
$$A + B = (A - B) \cup (B - A).$$
Prove each of the following.
a) $A + B = B + A$, b) $A + (B + C) = (A + B) + C$,
c) $A \cap (B + C) = (A \cap B) + (A \cap C)$, d) $A + A = \varnothing$, e) $A + \varnothing = A$.

9. Prove each of the following.
a) $A \cup B = \varnothing \Rightarrow A = \varnothing$ and $B = \varnothing$.
b) $A \cap B' = \varnothing$ if and only if $A \subseteq B$.
c) $A + B = \varnothing$ if and only if $A = B$.

10. Prove each of the following.
a) $A \cup C = B \cup C$ if and only if $A + B \subseteq C$.
b) $(A \cup C) + (B \cup C) = (A + B) - C$.

11. Use class algebra to prove that if $A \subseteq B$ and $C = B - A$, then $A = B - C$.

4 ORDERED PAIRS CARTESIAN PRODUCTS

If a is an element, we may use the axiom of class construction to form the class
$$\{a\} = \{x \mid x = a\}.$$
It is easy to see that $\{a\}$ contains only one element, namely the element a. A class containing a single element is called a *singleton*.

If a and b are elements, we may use the axiom of class construction to form the class
$$\{a, b\} = \{x \mid x = a \quad \text{or} \quad x = b\}.$$
Clearly $\{a, b\}$ contains two elements, namely the elements a and b. A class containing exactly two elements is called an *unordered pair*, or, more simply, a *doubleton*.

In like fashion, we can form the classes $\{a, b, c\}$, $\{a, b, c, d\}$, and so on.

Frequently, in mathematics, we need to form classes whose elements are doubletons. In order to be able to do this legitimately, we need a new axiom which will guarantee that if a and b are elements, then the doubleton $\{a, b\}$ is an element. This motivates our next axiom, which is often called the *Axiom of Pairing*:

A3. If a and b are elements, then $\{a, b\}$ is an element.

It is clear that $\{a, a\} = \{a\}$; thus, setting $a = b$ in Axiom A3, we immediately get

if a is an element, then the singleton $\{a\}$ is an element.

1.27 Theorem If $\{x, y\} = \{u, v\}$, then

$$[x = u \quad \text{and} \quad y = v] \qquad \text{or} \qquad [x = v \quad \text{and} \quad y = u].$$

The proof is left as an exercise for the reader. [*Hint*: Consider the cases $x = y$, $x \neq y$, separately. Use Axiom A1.]

An important notion in mathematics is that of an *ordered pair* of elements. Intuitively, an ordered pair is a class consisting of two elements in a specified order. In fact, the order is not really essential; what is essential is that ordered pairs have the following property.

1.28 Let (a, b) and (c, d) be ordered pairs. If $(a, b) = (c, d)$, then $a = c$ and $b = d$.

We would like to define ordered pairs in such a way as to avoid introducing a new undefined notion of "order." It is an interesting fact that this can, indeed, be accomplished; we proceed as follows.

1.29 Definition Let a and b be elements; the *ordered pair* (a, b) is defined to be the class

$$(a, b) = \{\{a\}, \{a, b\}\}.$$

By Axiom A3, (a, b) can be legitimately formed, *and is itself an element*.

It is worth noting that

$$(b, a) = \{\{b\}, \{b, a\}\} = \{\{b\}, \{a, b\}\}.$$

Hence there is a clear distinction between the two possible "orders" (a, b) and (b, a): they are distinct classes. It remains to prove that ordered pairs, as we have just defined them, have Property 1.28.

1.30 Theorem If $(a, b) = (c, d)$, then $a = c$ and $b = d$.

Proof. Suppose that $(a, b) = (c, d)$; that is,

$$\{\{a\}, \{a, b\}\} = \{\{c\}, \{c, d\}\}.$$

By Theorem 1.27, either

$$[\{a\} = \{c\} \quad \text{and} \quad \{a, b\} = \{c, d\}],$$

or

$$[\{a\} = \{c, d\} \quad \text{and} \quad \{a, b\} = \{c\}];$$

we will consider these two cases separately.

Case 1. $\{a\} = \{c\}$ and $\{a, b\} = \{c, d\}$. From $\{a\} = \{c\}$, it follows that $a = c$. From $\{a, b\} = \{c, d\}$ and Theorem 1.27, it follows that either $a = c$ and $b = d$, or $a = d$ and $b = c$; in the first case, we are done; in the second case, we have $b = c = a = d$, so again we are done.

Case 2. $\{a\} = \{c, d\}$ and $\{a, b\} = \{c\}$. Here $c \in \{c, d\}$ and $\{c, d\} = \{a\}$, so $c \in \{a\}$; thus $c = a$; analogously, $d = a$. Also, $b \in \{a, b\}$ and $\{a, b\} = \{c\}$, so $b \in \{c\}$; hence $b = c$. Thus $a = b = c = d$, and we are done. ■

1.31 Definition The *Cartesian product* of two classes A and B is the class of all ordered pairs (x, y) where $x \in A$ and $y \in B$. In symbols,

$$A \times B = \{(x, y) \mid x \in A \quad \text{and} \quad y \in B\}.$$

The following are a few simple properties of Cartesian products.

1.32 Theorem For all classes A, B, and C,

i) $A \times (B \cap C) = (A \times B) \cap (A \times C)$.

ii) $A \times (B \cup C) = (A \times B) \cup (A \times C)$.

iii) $(A \times B) \cap (C \times D) = (A \cap C) \times (B \cap D)$.

Proof

i)
$$\begin{aligned} (x, y) \in A \times (B \cap C) &\Leftrightarrow x \in A \quad \text{and} \quad y \in B \cap C \\ &\Leftrightarrow x \in A \quad \text{and} \quad y \in B \quad \text{and} \quad y \in C \\ &\Leftrightarrow (x, y) \in A \times B \quad \text{and} \quad (x, y) \in A \times C \\ &\Leftrightarrow (x, y) \in (A \times B) \cap (A \times C). \end{aligned}$$

iii)
$$\begin{aligned} (x, y) \in (A \times B) \cap (C \times D) &\Leftrightarrow (x, y) \in A \times B \quad \text{and} \quad (x, y) \in C \times D \\ &\Leftrightarrow x \in A \quad \text{and} \quad y \in B \quad \text{and} \quad x \in C \quad \text{and} \quad y \in D \\ &\Leftrightarrow x \in A \cap C \quad \text{and} \quad y \in B \cap D \\ &\Leftrightarrow (x, y) \in (A \cap C) \times (B \cap D). \text{ ■} \end{aligned}$$

Just as we found it instructive to represent relations between classes by means of Venn diagrams, it is often convenient to illustrate relations between products of classes by using a graphic device known as a *coordinate diagram.* A coordinate diagram is analogous to the familiar Cartesian coordinate plane; there are two axes—a vertical one and a horizontal one—but we consider only one "quadrant." If we wish to represent a class $A \times B$, then a segment of the horizontal axis is marked off to represent A and a segment of the vertical axis is marked off to represent B; $A \times B$ is the rectangle determined by these two segments (Fig. 5). As an example of the use of coordinate diagrams, Theorem 1.32(iii) is illustrated in Fig. 6.

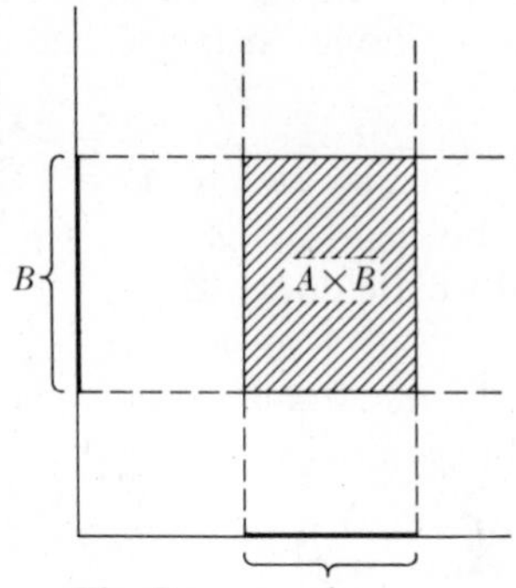

Fig. 5

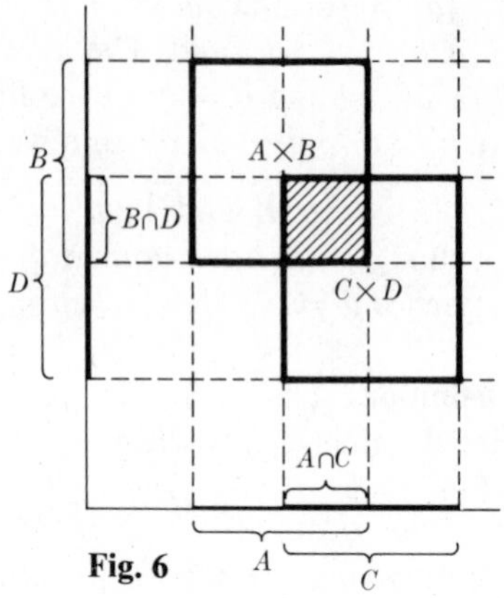

Fig. 6

EXERCISES 1.4

1. Let $A = \{a, b, c, d\}$, $B = \{1, 2, 3\}$, $C = \{x, y, z\}$. Find $A \times B$, $B \times A$, $C \times (B \times A)$, $(A \cup B) \times C$, $(A \times C) \cup (B \times C)$, $(A \cup B) \times (B \cup C)$.
2. Prove Theorem 1.32(ii).
3. Prove that $A \times (B - D) = (A \times B) - (A \times D)$.
4. Prove that $(A \times B) \cap (C \times D) = (A \times D) \cap (C \times B)$.
5. If A, B and C are classes, prove the following.
   a) $(A \times A) \cap (B \times C) = (A \cap B) \times (A \cap C)$.
   b) $(A \times B) - (C \times C) = [(A - C) \times B] \cup [A \times (B - C)]$.
   c) $(A \times A) - (B \times C) = [(A - B) \times A] \cup [A \times (A - C)]$.
6. Prove that A and B are disjoint if and only if, for any nonempty class C, $A \times C$ and $B \times C$ are disjoint.
7. If A and C are nonempty classes, prove that $A \subseteq B$ and $C \subseteq D$ if and only if $A \times C \subseteq B \times D$.
8. Let A, B, C, D be nonempty classes. Prove that $A \times B = C \times D$ if and only if $A = C$ and $B = D$.
9. If A, B, and C are any classes, prove
   a) $A \times B$ and $A' \times C$ are disjoint, b) $B \times A$ and $C \times A'$ are disjoint.
10. Prove that $A \times B = \varnothing$ if and only if $A = \varnothing$ or $B = \varnothing$.

11. Prove each of the following.
    a) If $a = \{b\}$, then $b \in a$.
    b) $x = y$ if and only if $\{x\} = \{y\}$.
    c) $x \in a$ if and only if $\{x\} \subseteq a$.
    d) $\{a, b\} = \{a\}$ if and only if $a = b$.
12. We give the following alternative definition of ordered pairs: $(x, y) = \{\{x, \varnothing\}, \{y, \{\varnothing\}\}\}$. Using this definition, prove that
$$(a, b) = (c, d) \Rightarrow a = c \text{ and } b = d.$$

5 GRAPHS

A class of ordered pairs is called a *graph.* In other words, a graph is an arbitrary subclass of $\mathscr{U} \times \mathscr{U}$.

The importance of graphs will become apparent to the reader in Chapters 2 and 3. It may be shown, for instance, that a function from A to B is a graph $G \subseteq A \times B$ with certain special properties. Specifically, G consists of all the pairs (x, y) such that $y = f(x)$. This example may help to motivate the following definitions.

1.33 Definition If G is a graph, then G^{-1} is the graph defined by
$$G^{-1} = \{(x, y) \mid (y, x) \in G\}.$$

1.34 Definition If G and H are graphs, then $G \circ H$ is the graph defined as follows:
$$G \circ H = \{(x, y) \mid \exists z \ni (x, z) \in H \quad \text{and} \quad (z, y) \in G\}.$$

The following are a few basic properties of graphs.

1.35 Theorem If G, H, and J are graphs, then the following statements hold:
i) $(G \circ H) \circ J = G \circ (H \circ J)$.
ii) $(G^{-1})^{-1} = G$.
iii) $(G \circ H)^{-1} = H^{-1} \circ G^{-1}$.

Proof

i)
$$\begin{aligned}(x, y) \in (G \circ H) \circ J &\Leftrightarrow \exists z \ni (x, z) \in J \quad \text{and} \quad (z, y) \in G \circ H \\ &\Leftrightarrow \exists w \text{ and } \exists z \ni (x, z) \in J \text{ and } (z, w) \in H \text{ and } (w, y) \in G \\ &\Leftrightarrow \exists w \ni (x, w) \in H \circ J \quad \text{and} \quad (w, y) \in G \\ &\Leftrightarrow (x, y) \in G \circ (H \circ J).\end{aligned}$$

ii) $(x, y) \in (G^{-1})^{-1} \Leftrightarrow (y, x) \in G^{-1}$

$$\Leftrightarrow (x, y) \in G.$$

iii) $(x, y) \in (G \circ H)^{-1} \Leftrightarrow (y, x) \in G \circ H$

$$\Leftrightarrow \exists z \ni (y, z) \in H \quad \text{and} \quad (z, x) \in G$$
$$\Leftrightarrow \exists z \ni (x, z) \in G^{-1} \quad \text{and} \quad (z, y) \in H^{-1}$$
$$\Leftrightarrow (x, y) \in H^{-1} \circ G^{-1}. \blacksquare$$

1.36 Definition Let G be a graph. By the *domain* of G we mean the class

$$\text{dom } G = \{x \mid \exists y \ni (x, y) \in G\},$$

and by the *range* of G we mean the class

$$\text{ran } G = \{y \mid \exists x \ni (x, y) \in G\}.$$

In other words, the domain of G is the class of all "first components" of elements of G, and the range of G is the class of all "second components" of elements of G.

1.37 Theorem If G and H are graphs, then

i) $\text{dom } G = \text{ran } G^{-1}$, ii) $\text{ran } G = \text{dom } G^{-1}$,

iii) $\text{dom } (G \circ H) \subseteq \text{dom } H$, iv) $\text{ran } (G \circ H) \subseteq \text{ran } G$.

Proof

i) $x \in \text{dom } G \Leftrightarrow \exists y \ni (x, y) \in G$

$$\Leftrightarrow \exists y \ni (y, x) \in G^{-1}$$
$$\Leftrightarrow x \in \text{ran } G^{-1}.$$

iii) $x \in \text{dom } (G \circ H) \Rightarrow \exists y \ni (x, y) \in (G \circ H)$

$$\Rightarrow \exists z \ni (x, z) \in H \quad \text{and} \quad (z, y) \in G$$
$$\Rightarrow x \in \text{dom } H. \blacksquare$$

1.38 Corollary Let G and H be graphs. If $\text{ran } H \subseteq \text{dom } G$ then $\text{dom } G \circ H = \text{dom } H$.

The proof of this theorem is left as an exercise for the reader.

EXERCISES 1.5

1. Let

$$G = \{(b, b), (b, c), (c, c), (c, d)\}$$

and

$$H = \{(b, a), (c, b), (d, c)\}.$$

Find $G^{-1}, H^{-1}, G \circ H, H \circ G, (G \circ H)^{-1}, (G \cup H)^{-1}, H^{-1} \circ G$.

2. Prove Theorem 1.37, parts (ii) and (iv).

3. Prove Theorem 1.38.

4. If G, H, and J are graphs, prove each of the following.

   a) $(H \cup J) \circ G = (H \circ G) \cup (J \circ G)$,
   b) $(G - H)^{-1} = G^{-1} - H^{-1}$,
   c) $G \circ (H \cap J) \subseteq (G \circ H) \cap (G \circ J)$,
   d) $(G \circ H) - (G \circ J) \subseteq G \circ (H - J)$.

5. If G and H are graphs, prove each of the following.

   a) $(G \cap H)^{-1} = G^{-1} \cap H^{-1}$,
   b) $(G \cup H)^{-1} = G^{-1} \cup H^{-1}$.

6. If G, H, J, and K are graphs, prove

   a) if $G \subseteq H$ and $J \subseteq K$, then $G \circ J \subseteq H \circ K$,
   b) $G \subseteq H$ if and only if $G^{-1} \subseteq H^{-1}$.

7. If A, B, and C are classes, prove each of the following.

   a) $(A \times B)^{-1} = B \times A$.
   b) If $A \cap B \neq \varnothing$, then $(A \times B) \circ (A \times B) = A \times B$.
   c) If A and B are disjoint, then $(A \times B) \circ (A \times B) = \varnothing$.
   d) If $B \neq \varnothing$, then $(B \times C) \circ (A \times B) = A \times C$.

8. Let G and H be graphs; prove each of the following.

   a) If $G \subseteq A \times B$, then $G^{-1} \subseteq B \times A$.
   b) If $G \subseteq A \times B$ and $H \subseteq B \times C$, then $H \circ G \subseteq A \times C$.

9. If G and H are graphs, prove each of the following.

   a) $\operatorname{dom}(G \cup H) = (\operatorname{dom} G) \cup (\operatorname{dom} H)$.
   b) $\operatorname{ran}(G \cup H) = (\operatorname{ran} G) \cup (\operatorname{ran} H)$.
   c) $\operatorname{dom} G - \operatorname{dom} H \subseteq \operatorname{dom}(G - H)$.
   d) $\operatorname{ran} G - \operatorname{ran} H \subseteq \operatorname{ran}(G - H)$.

10. Let G be a graph, and let B be a subclass of the domain of G. By the *restriction of G to B* we mean the graph

$$G_{[B]} = \{(x, y) \mid (x, y) \in G \text{ and } x \in B\}.$$

Prove each of the following.

   a) $G_{[B]} = G \cap (B \times \operatorname{ran} G)$,
   b) $G_{[B \cup C]} = G_{[B]} \cup G_{[C]}$,
   c) $G_{[B \cap C]} = G_{[B]} \cap G_{[C]}$,
   d) $(G \circ H)_{[B]} = G \circ H_{[B]}$.

11. Let G be a graph and let B be a subclass of the domain of G. We use the symbol $G(B)$ to designate the class

$$G(B) = \{y \mid \exists x \in B \ni (x, y) \in G\}.$$

Prove each of the following.

   a) $G(B) = \operatorname{ran} G_{[B]}$,
   b) $G(B \cup C) = G(B) \cup G(C)$,
   c) $G(B \cap C) = G(B) \cap G(C)$,
   d) If $B \subseteq C$, then $G(B) \subseteq G(C)$.

6 GENERALIZED UNION AND INTERSECTION

Consider the class $\{A_1, A_2, \ldots, A_n\}$; its elements are indexed by the numbers 1, 2, ..., n. Such a class if often called an indexed family of classes; the numbers 1, 2, ..., n are called indices and the class $\{1, 2, \ldots, n\}$ is called the index class.

More generally, we are frequently led to think of a class I whose elements $i, j, k, \ldots$ serve as indices to designate the elements of a class $\{A_i, A_j, A_k, \ldots\}$. The class $\{A_i, A_j, A_k, \ldots\}$ is called an *indexed family of classes*, I is called its *index class*, and the elements of I are called *indices*. A compact notation which is often used to designate the class $\{A_i, A_j, A_k, \ldots\}$ is

$$\{A_i\}_{i \in I}.$$

Thus, speaking informally, $\{A_i\}_{i \in I}$ is the class of all the classes A_i, as i ranges over I.

Remark. The definition of an indexed family of classes which we have just given is, admittedly, an intuitive one; it relies on the intuitive notion of *indexing*. This intuitive definition is adequate at the present time; however, for future reference, we now give a *formal* definition of the same concept:

By an indexed family of classes, $\{A_i\}_{i \in I}$, we mean a graph G whose domain is I; for each $i \in I$ we define A_i by

$$A_i = \{x \mid (i, x) \in G\}.$$

For example, consider $\{A_i\}_{i \in I}$ where $I = \{1, 2\}$, $A_1 = \{a, b\}$, and $A_2 = \{c, d\}$. Then, *formally*, $\{A_i\}_{i \in I}$ is the graph

$$G = \{(1, a), (1, b), (2, c), (2, d)\}.$$

If $\{A_i\}_{i \in I}$ is an indexed family of classes such that for each $i \in I$, A_i is an element, then we let $\{A_i \mid i \in I\}$ designate the class whose elements are all the A_i, that is, $\{A_i \mid i \in I\} = \{x \mid x = A_i \text{ for some } i \in I\}$. However, we shall follow current mathematical usage and use the two expressions, $\{A_i\}_{i \in I}$ and $\{A_i \mid i \in I\}$, interchangeably.

1.39 Definition Let $\{A_i\}_{i \in I}$ be an indexed family of classes. The *union of the classes* A_i consists of all the elements which belong to at least one class A_i of the family. In symbols,

$$\bigcup_{i \in I} A_i = \{x \mid \exists j \in I \ni x \in A_j\}.$$

The *intersection of the classes* A_i consists of all the elements which belong to every class A_i of the family. In symbols,

$$\bigcap_{i \in I} A_i = \{x \mid \forall i \in I, x \in A_i\}.$$

The following are some basic properties of indexed families of classes.

1.40 Theorem Let $\{A_i\}_{i\in I}$ be an indexed family of classes.

i) If $A_i \subseteq B$ for every $i \in I$, then $\bigcap_{i\in I} A_i \subseteq B$.

ii) If $B \subseteq A_i$ for every $i \in I$, then $B \subseteq \bigcap_{i\in I} A_i$.

Proof

i) Suppose that $A_i \subseteq B$ for every $i \in I$; now if $x \in \bigcup_{i\in I} A_i$, then $x \in A_j$ for some $j \in I$; but $A_j \subseteq B$, so $x \in B$. Thus $\bigcup_{i\in I} A_i \subseteq B$.

The proof of (ii) is left as an exercise for the reader. ■

1.41 Theorem (*Generalized deMorgan's Laws*). Let $\{A_i\}_{i\in I}$ be an index family of classes. Then,

i) $(\bigcup_{i\in I} A_i)' = \bigcap_{i\in I} A_i'$.

ii) $(\bigcap_{i\in I} A_i)' = \bigcup_{i\in I} A_i'$.

Proof

i)
$$\begin{aligned} x \in (\bigcup_{i\in I} A_i)' &\Leftrightarrow x \notin \bigcup_{i\in I} A_i \\ &\Leftrightarrow \forall i \in I, x \notin A_i \\ &\Leftrightarrow \forall i \in I, x \in A_i' \\ &\Leftrightarrow x \in \bigcap_{i\in I} A_i'. \end{aligned}$$

The proof of (ii) is left as an exercise for the reader. ■

1.42 Theorem (*Generalized Distributive Laws*). Let $\{A_i\}_{i\in I}$ and $\{B_j\}_{j\in J}$ be indexed families of classes. Then

i) $(\bigcup_{i\in I} A_i) \cap (\bigcup_{j\in J} B_j) = \bigcup_{(i,j)\in I\times J} (A_i \cap B_j)$,

ii) $(\bigcap_{i\in I} A_i) \cup (\bigcap_{j\in J} B_j) = \bigcap_{(i,j)\in I\times J} (A_i \cup B_j)$.

Proof

i)
$$\begin{aligned} x \in (\bigcup_{i\in I} A_i) \cap (\bigcup_{j\in J} B_j) &\Leftrightarrow x \in \bigcup_{i\in I} A_i \quad \text{and} \quad x \in \bigcup_{j\in J} B_j \\ &\Leftrightarrow x \in A_h \text{ for some } h \in I \text{ and } x \in B_k \text{ for some } k \in J \\ &\Leftrightarrow x \in A_h \cap B_k \text{ for some } (h, k) \in I \times J \\ &\Leftrightarrow x \in \bigcup_{(i,j)\in I\times J} (A_i \cap B_j). \end{aligned}$$

The proof of (ii) is left as an exercise for the reader. ■

A theorem concerning the union of graphs will be useful to us in the next chapter.

1.43 Theorem Let $\{G_i\}_{i\in I}$ be a family of graphs. Then

i) $\operatorname{dom}\left(\bigcup_{i\in I} G_i\right) = \bigcup_{i\in I} (\operatorname{dom} G_i)$.

ii) $\operatorname{ran}\left(\bigcup_{i\in I} G_i\right) = \bigcup_{i\in I} (\operatorname{ran} G_i)$.

Proof

i)
$$\begin{aligned} x \in \operatorname{dom}\left(\bigcup_{i\in I} G_i\right) &\Leftrightarrow \exists y \ni (x, y) \in \bigcup_{i\in I} G_i && \text{by 1.36} \\ &\Leftrightarrow \exists y \ni (x, y) \in G_j \text{ for some } j \in I && \text{by 1.39} \\ &\Leftrightarrow x \in \operatorname{dom} G_j \text{ for some } j \in I && \text{by 1.36} \\ &\Leftrightarrow x \in \bigcup_{i\in I} (\operatorname{dom} G_i) && \text{by 1.39.} \end{aligned}$$

The proof of (ii) is left as an exercise for the reader. ■

A variant notation for the union and intersection of a family of classes is sometimes useful. If $\mathscr{A}$ is a class (its elements are necessarily classes), we define the *union of* $\mathscr{A}$, or *union of the elements of* $\mathscr{A}$, to be the union of all the classes which are elements of $\mathscr{A}$. In symbols,

1.44
$$\bigcup_{A\in\mathscr{A}} A = \{x \mid x \in A \text{ for some } A \in \mathscr{A}\}.$$

In other words, $x \in \bigcup_{A\in\mathscr{A}} A$ if and only if there is a class A such that $x \in A$ and $A \in \mathscr{A}$. Analogously, we define the *intersection of* $\mathscr{A}$, or *intersection of the elements of* $\mathscr{A}$, to be the intersection of all the classes which are elements of $\mathscr{A}$. In symbols,

1.45
$$\bigcap_{A\in\mathscr{A}} A = \{x \mid x \in A \text{ for every } A \in \mathscr{A}\}.$$

1.46 Example Let $\mathscr{A} = \{K, L, M\}$, where $K = \{a, b, d\}$, $L = \{a, c, d\}$, and $M = \{d, e\}$. Then

$$\bigcup_{A\in\mathscr{A}} A = \{a, b, c, d, e\} \qquad \text{and} \qquad \bigcap_{A\in\mathscr{A}} A = \{d\}.$$

1.47 *Remark.* It is frequent practice, in the literature of set theory, to write

$$\bigcup \mathscr{A} \quad \text{for} \quad \bigcup_{A\in\mathscr{A}} A$$

and

$$\bigcap \mathscr{A} \quad \text{for} \quad \bigcap_{A \in \mathscr{A}} A.$$

We shall occasionally follow that practice in this book.

EXERCISES 1.6

1. Prove Theorem 1.40(ii).
2. Prove Theorem 1.41(ii).
3. Prove Theorem 1.42(ii).
4. Prove Theorem 1.43(ii).
5. Let $\{A_i\}_{i \in I}$ and $\{B_i\}_{i \in I}$ be two families of classes with the same index class I. Suppose that $\forall i \in I$, $A_i \subseteq B_i$; prove that

   a) $\bigcup_{i \in I} A_i \subseteq \bigcup_{i \in I} B_i$, b) $\bigcap_{i \in I} A_i \subseteq \bigcap_{i \in I} B_i$.
6. Let $\{A_i\}_{i \in I}$ and $\{B_j\}_{j \in J}$ be indexed families of classes. Prove the following.

   a) $(\bigcap_{i \in I} A_i) \times (\bigcap_{j \in J} B_j) = \bigcap_{(i,j) \in I \times J} (A_i \times B_j)$,

   b) $(\bigcup_{i \in I} A_i) \times (\bigcup_{j \in J} B_j) = \bigcup_{(i,j) \in I \times J} (A_i \times B_j)$.
7. Let $\{A_i\}_{i \in I}$ and $\{B_j\}_{j \in J}$ be indexed families of classes. Suppose that $\forall i \in I$, $\exists j \in J \ni B_j \subseteq A_i$. Prove that
$$\bigcap_{j \in J} B_j \subseteq \bigcap_{i \in I} A_i.$$
8. Let $\{A_i\}_{i \in I}$ and $\{B_j\}_{j \in J}$ be indexed families of classes. Prove that

   a) $(\bigcup_{i \in I} A_i) - (\bigcup_{j \in J} B_j) = \bigcup_{i \in I} (\bigcap_{j \in J} [A_i - B_j])$,

   b) $(\bigcap_{i \in I} A_i) - (\bigcap_{j \in J} B_j) = \bigcap_{i \in I} (\bigcup_{j \in J} [A_i - B_j])$.
9. We say that an indexed family $\{B_i\}_{i \in I}$ is a *covering* of A if $A \subseteq \bigcup_{i \in I} B_i$. Suppose that $\{B_i\}_{i \in I}$ and $\{C_j\}_{j \in J}$ are two distinct coverings of A. Prove that the family $\{(B_i \cap C_j)\}_{(i,j) \in I \times J}$ is a covering of A.
10. Let $a = \{u, v, w\}$, $b = \{w, x\}$, $c = \{w, y\}$, $r = \{a, b\}$, $s = \{b, c\}$, and $p = \{r, s\}$. Find the classes $\cup(\cup p)$, $\cap(\cap p)$, $\cup(\cap p)$, $\cap(\cup p)$.
11. Prove that $\cap(\mathscr{A} \cup \mathscr{B}) = (\cap\mathscr{A}) \cap (\cap\mathscr{B})$.
12. Prove each of the following.

    a) If $A \in \mathscr{B}$, then $A \subseteq \cup \mathscr{B}$ and $\cap\mathscr{B} \subseteq A$.

    b) $\mathscr{A} \subseteq \mathscr{B}$ if and only if $\cup\mathscr{A} \subseteq \cup\mathscr{B}$.

    c) If $\varnothing \in \mathscr{A}$, then $\cap\mathscr{A} = \varnothing$.

7 SETS

Undoubtedly, everything we have said in the preceding pages is fairly familiar to the reader. Even though we said *class* where the reader is more accustomed to hearing *set*, it is obvious that the "union" and "intersection" defined in this chapter are precisely the familiar union and intersection of sets, the "Cartesian product" is exactly the usual Cartesian product of sets, and similarly for the other concepts introduced in this chapter. At this point, it appears as though everything we are accustomed to doing with sets can be done with classes. Thus, the reader may very well ask, "Why bother distinguishing between classes and sets? Why not develop all of mathematics in terms of classes?" Since, as we have said, a class means "any collection of elements," why not simply call a class a set, and be done with it? The answer to this question is of great importance; the chief purpose of this section is to explain why we *do* want to distinguish between classes and sets.

First, we note that the axiom of class construction (Axiom A2) permits us to form the *class of all elements* which satisfy a given property; it does not allow us to form the *class of all classes* which satisfy a given property. The reason for this limitation is obvious: if we were enabled to form the class of all classes which satisfy any given property, then we could form "Russell's class" of all classes which are not elements of themselves, and this would give us Russell's paradox.

Next, we note that in mathematics we often need to form particular *sets of sets.* A few examples which come to mind are the following:

The set of all closed intervals $[a, b]$ of real numbers.

The set of all convergent sequences of real numbers.

The set of all the lines in the plane (where each line is regarded as a set of points).

Let us look more closely at the first example; in a discussion in elementary calculus, we would feel perfectly free to say "let $\mathscr{A}$ consist of all the sets which are closed intervals $[a, b]$ of real numbers." Now if "sets" were no different from "classes" then, by the preceding paragraph, we would not be allowed to form $\mathscr{A}$. This would be an intolerable restriction upon our freedom of operating with sets in mathematics.

Let us recapitulate: The notion of *class* is appealing because of its intuitive simplicity and generality; however, there is a serious drawback to dealing with classes, namely that it is not permissible [for an arbitrary property $P(X)$] to form the "class of all classes X which satisfy $P(X)$." This would be an intolerable restriction on our mathematical freedom of action if we were to base mathematics upon classes. Instead, we base mathematics upon *sets*; the concept of a set is somewhat narrower than that of a class; sets will be defined in such a way, however, that for any property $P(X)$, *it is legitimate to form the class of all sets X which satisfy* $P(X)$. Thus, the freedom we require is restored to us, provided we are willing to deal with sets rather than the broader notion of classes. We are more than willing to do so, because, as we will be able to show, all the classes we deal with in mathematics *are sets.*

Now, how should sets be defined? In order to answer this question, it is essential to remind ourselves, once again, that we are seeking a way of defining sets which makes it legitimate to form the class of all *sets* X which satisfy any property $P(X)$. One obvious answer is to identify sets with elements: let a set be the same thing as an element. Then Axiom A2 will certainly permit us to form the class of all sets X which satisfy any property $P(X)$.

This simple answer is, in fact, the one which has been adopted by most mathematicians. We will use it here; thus,

1.48 Definition By a *set* is meant any class which is an element of a class.

This definition is supported by our intuitive perception of what a set should be. For if $A, B, C, \ldots$ are sets, it is perfectly reasonable that we should be able to form the class $\{A, B, C, \ldots\}$ whose elements are $A, B, C, \ldots$. In other words, we would quite certainly expect every set to be an element. The converse is equally reasonable: for if A is *not* a set, then A is a proper class, and we have already seen that in order to avoid contradictions, proper classes should not be elements of anything. Thus, if A is not a set, then A is not an element.

In the remainder of this section, we will state the basic axioms dealing with sets. The main purpose of these axioms is to guarantee that when the usual set-theoretic operations are performed on sets, the result, each time, is a set.

First, we note that the Axiom of Pairing, our Axiom A3, may be re-stated thus:

A3. If a and b are sets, then $\{a, b\}$ is a set.

Now, if B is a set and $A \subseteq B$, one would reasonably expect A to be a set. This is the content of our next axiom, called the *Axiom of Subsets.*

A4. Every subclass of a set is a set.

By Theorem 1.20(ii), $A \cap B \subseteq A$; thus, by Axiom A4, if A is a set, then $A \cap B$ is a set. In particular, *the intersection of any two sets is a set.*

The union of "not too many" sets should be a set. This is guaranteed by our next axiom, called the *Axiom of Unions*:

A5. If $\mathscr{A}$ is a set of sets, then $\bigcup_{A \in \mathscr{A}} A$ is a set.

If A and B are sets, then, by Axiom A3, $\{A, B\}$ is a set; it follows immediately from Definition 1.44 that $\bigcup_{X \in \{A, B\}} X = A \cup B$; thus, by Axiom A5, $A \cup B$ is a set. This shows that *the union of two sets is a set.*

1.49 *Remark.* By Axiom A3, every doubleton is a set. Furthermore, letting $a = b$ in A3, every singleton is a set. Since the union of two sets is a set, it follows that every class of three elements is a set, every class of four elements is a set, and so on. Thus, in an intuitive sense, every finite class is a set.

Next, we will establish that if A is a set, then the class of all the subsets of A is a set. We begin with a definition.

1.50 Definition Let A be a set; by the *power set* of A we mean the class of all the subsets of A. In symbols, the power set of A is the class

$$\mathscr{P}(A) = \{B \mid B \subseteq A\}.$$

Note that by Axiom A4, $\mathscr{P}(A)$ is the class of all the *sets* B which satisfy $B \subseteq A$. By Definition 1.48 and Axiom A2, it is legitimate to form this class.

The following is called the *Axiom of Power Sets*:

A6. If A is a set, then $\mathscr{P}(A)$ is a set.

1.51 Example If $A = \{a, b\}$, then $\mathscr{P}(A) = \{\varnothing, \{a\}, \{b\}, \{a, b\}\}$.

From all that we have said so far, it does not yet follow that there *exist* any sets at all. To fill this vacuum, we state a temporary axiom, which will be superseded by Axiom A9:

T The empty class is a set.

Henceforth, we will refer to $\varnothing$ as the empty *set*.

From Axiom T, together with A3 and A5, we may infer the existence of a great many sets. We have the empty set, $\varnothing$; we have singletons such as $\{\varnothing\}$, $\{\{\varnothing\}\}$, etc.; we have doubletons such as $\{\varnothing, \{\varnothing\}\}$, formed by any two of the above. Similarly, taking unions of the above repeatedly, we may form sets with *any finite number* of elements.

1.52 *Remark.* An important consequence of Axiom A6 is the following. If A is a set, then clearly

$$B = \{X \mid X \subseteq A \quad \text{and} \quad P(X)\}$$

is the class of all the subsets of A which satisfy the property P; by Axiom A4, Axiom A2 may legitimately be used to form the class B. Now if $X \in B$, then X is a subset of A, so $X \in \mathscr{P}(A)$; thus $B \subseteq \mathscr{P}(A)$. But by Axiom A6, $\mathscr{P}(A)$ is a set, hence by Axiom A4, B is a set. We may summarize as follows: if A is a set and $P(X)$ is a property of X, then *the class of all the subsets of A which satisfy $P(X)$ is a set.*

1.53 Theorem If A and B are sets, then $A \times B$ is a set.

Proof. Let A and B be sets. By Axiom A5, $A \cup B$ is a set; by Axiom A6, $\mathscr{P}(A \cup B)$ is a set; finally, by Axiom A6 again, $\mathscr{P}[\mathscr{P}(A \cup B)]$ is a set. We will prove that $A \times B \subseteq \mathscr{P}[\mathscr{P}(A \cup B)]$, and it will follow, by Axiom A4, that $A \times B$ is a set.

Let $(x, y) \in A \times B$. By 1.29, $(x, y) = \{\{x\}, \{x, y\}\}$. Now $x \in A \cup B$, hence $\{x\} \subseteq A \cup B$, so $\{x\} \in \mathscr{P}(A \cup B)$. Similarly, $x \in A \cup B$ and $y \in A \cup B$, so $\{x, y\} \subseteq A \cup B$, hence $\{x, y\} \in \mathscr{P}(A \cup B)$. We have just shown that $\{x\}$ and $\{x, y\}$ are elements of $\mathscr{P}(A \cup B)$, hence

$$\{\{x\}, \{x, y\}\} \subseteq \mathscr{P}(A \cup B);$$

it follows that

$$\{\{x\}, \{x, y\}\} \in \mathscr{P}[\mathscr{P}(A \cup B)],$$

that is,

$$(x, y) \in \mathscr{P}[\mathscr{P}(A \cup B)].$$

Thus

$$A \times B \subseteq \mathscr{P}[\mathscr{P}(A \cup B)]. \blacksquare$$

It follows from Theorem 1.53 and Axiom A4 that if A and B are sets, then any graph $G \subseteq A \times B$ is a set.

It is easy to show that if G is a set, then dom G and ran G are sets (see Exercise 5, Exercise Set 1.7). Using this fact, one can easily show that if G and H are sets, then $G \circ H$ and G^{-1} are sets (see Exercise 6, Exercise Set 1.7).

EXERCISES 1.7

1. If A and B are sets, prove that $A - B$ and $A + B$ are sets. (See Exercise 8, Exercise Set 1.3.)
2. If A is a proper class and $A \subseteq B$, prove that B is a proper class. Conclude that the union of two proper classes is a proper class.
3. Prove that the "Russell class" and the universal class are proper classes. [*Hint*. Use the result of Exercise 8, Exercise Set 1.2.]
4. Let $\{A_i\}_{i \in I}$ be an indexed family of sets. Prove that $\bigcap_{i \in I} A_i$ is a set.
5. Let G be a graph. Prove that if G is a set, then dom G and ran G are sets. [*Hint:* Show that both dom G and ran G are subsets of $\cup(\cup G)$.]

6. Let G and H be graphs. Prove that if G and H are sets, then G^{-1} and $G \circ H$ are sets.
7. Let $r = \{a, b\}$, $s = \{b, c\}$, $p = \{r, s\}$. Find the sets $\mathscr{P}(r)$, $\mathscr{P}(\mathscr{P}(r))$, and $\mathscr{P}(\cup p)$.
8. Let A and B be sets; prove the following.
   a) $A \subseteq B$ if and only if $\mathscr{P}(A) \subseteq \mathscr{P}(B)$.
   b) $A = B$ if and only if $\mathscr{P}(A) = \mathscr{P}(B)$.
   c) $\mathscr{P}(A) \cap \mathscr{P}(B) = \mathscr{P}(A \cap B)$.
   d) $\mathscr{P}(A) \cup \mathscr{P}(B) \subseteq \mathscr{P}(A \cup B)$.
   e) $A \cap B = \varnothing$ if and only if $\mathscr{P}(A) \cap \mathscr{P}(B) = \{\varnothing\}$.
9. If A and $\mathscr{B}$ are sets, prove the following.
   a) $\cup(\mathscr{P}(\mathscr{B})) = \mathscr{B}$.
   b) $\cap(\mathscr{P}(\mathscr{B})) = \varnothing$.
   c) If $\mathscr{P}(A) \in \mathscr{P}(\mathscr{B})$ then $A \in \mathscr{B}$.
10. Exhibit the sets $\mathscr{P}(\mathscr{P}(\varnothing))$ and $\mathscr{P}[\mathscr{P}(\mathscr{P}(\varnothing))]$.

2

Functions

1 INTRODUCTION

The concept of a function is one of the most basic mathematical ideas and enters into almost every mathematical discussion. A function is generally defined as follows: If A and B are classes, then a function from A to B is a rule which to every element $x \in A$ assigns a unique element $y \in B$; to indicate this connection between x and y we usually write $y = f(x)$. For instance, consider the function $y = \sin x$; if we take A to be the set of all the real numbers and B to be the closed interval $[-1, 1]$, then it is easy to see that $y = \sin x$ is a rule which, to every number $x \in A$, assigns a unique number $y \in B$.

The graph of a function is defined as follows: If f is a function from A to B, then the graph of f is the class of all ordered pairs (x, y) such that $y = f(x)$. For example, let $A = \{a, b, c\}$ and $B = \{d, e\}$, and let f be the function defined by the following table.

x	$f(x)$
a	d
b	e
c	d

The graph of f is $\{(a, d), (b, e), (c, d)\}$.

Clearly, we may use the information contained in the table to construct the graph of f; we may also operate the other way, that is, we may use the information contained in the graph to construct the table of f. Thus a function f completely determines its graph, and conversely, its graph completely determines f. Hence there is no need to distinguish between a function and its graph.

Since a function and its graph are essentially one and the same thing, we may, if we wish, *define* a function to be a graph. There is an important advantage to be gained by doing this—namely, we avoid having to introduce the word *rule* as a new undefined concept of set theory. For this reason it is customary, in rigorous treatments of mathematics, to introduce the notion of *function* via that of *graph*. We shall follow that procedure here.

2 FUNDAMENTAL CONCEPTS AND DEFINITIONS

We begin by giving our "official" definition of a function.

2.1 Definition A *function from A to B* is a triple* of objects $\langle f, A, B\rangle$, where A and B are classes and f is a subclass of $A \times B$ with the following properties.

F1. $\forall x \in A, \exists y \in B$ such that $(x, y) \in f$.

F2. If $(x, y_1) \in f$ and $(x, y_2) \in f$, then $y_1 = y_2$.

It is customary to write $f: A \to B$ instead of $\langle f, A, B\rangle$.

In ordinary mathematical applications, every function $f: A \to B$ is a function from a *set* A to a *set* B. However, the intuitive concept of a function from A to B is meaningful for any two collections A and B, whether A and B be sets or proper classes; hence it is natural to give the definition of a function in its most general form, letting A and B be any classes. Once again, every set is a class, hence everything we have to say about functions from a class A to a class B applies, in particular, to functions from a set A to a set B.

Let $f: A \to B$ be a function; if $(x, y) \in f$, we say that *y is the image of x* (with respect to f); we also say that *x is the pre-image of y* (with respect to f); we also say that *f maps x onto y*, and symbolize this statement by $x \overset{f}{\mapsto} y$. (The reader may, if he wishes, picture these statements as in Fig. 1.)

Thus, F1 states that

every element $x \in A$ has an image $y \in B$.

F2 states that if $x \in A$, then

the image of x is unique;

for if $(x, y_1) \in f$ and $(x, y_2) \in f$, that is, if y_1 and y_2 are both images of x, then F2 dictates that $y_1 = y_2$. It follows that F1 and F2 combined state that

2.2 Every element $x \in A$ has a uniquely determined image $y \in B$.

2.3 Theorem Let A and B be classes and let f be a graph. Then $f: A \to B$ is a function if and only if

i) F2 holds, ii) $\operatorname{dom} f = A$, and iii) $\operatorname{ran} f \subseteq B$.

Proof. Suppose $f: A \to B$ is a function; by 2.1, F2 holds. Furthermore,

---

* If f, A and B are not all sets, the ordered triple $\langle f, A, B\rangle$ can be defined, formally, thus:

$$\langle f, A, B\rangle = (f \times \{\varnothing\}) \cup (A \times \{\{\varnothing\}\}) \cup (B \times \{\{\{\varnothing\}\}\}).$$

a) $x \in \operatorname{dom} f \Rightarrow \exists y \ni (x, y) \in f$	by 1.36
$\Rightarrow (x, y) \in A \times B$	because $f \subseteq A \times B$ by 2.1
$\Rightarrow x \in A$	by 1.31.
b) $x \in A \Rightarrow \exists y \in B \ni (x, y) \in f$	by F1
$\Rightarrow x \in \operatorname{dom} f$	by 1.36.
c) $y \in \operatorname{ran} f \Rightarrow \exists x \ni (x, y) \in f$	by 1.36
$\Rightarrow (x, y) \in A \times B$	because $f \subseteq A \times B$ by 2.1
$\Rightarrow y \in B$	by 1.31.

By (a) and (b), $\operatorname{dom} f = A$; by (c), $\operatorname{ran} f \subseteq B$. Thus, (i), (ii), and (iii) hold.

For the converse, suppose that (i), (ii), and (iii) hold.

a) $(x, y) \in f \Rightarrow x \in \operatorname{dom} f$ and $y \in \operatorname{ran} f$	by 1.36
$\Rightarrow x \in A$ and $y \in B$	by (ii) and (iii)
$\Rightarrow (x, y) \in A \times B$	by 1.31.

Thus, $f \subseteq A \times B$.

b) Let x be an arbitrary element of A. By (ii), $x \in \operatorname{dom} f$; hence $\exists y \ni (x, y) \in f$; but $y \in \operatorname{ran} f$, so by (iii), $y \in B$. This proves that F1 holds. By (i), F2 holds; thus, by 2.1, $f: A \to B$ is a function. ■

From Theorem 2.3 we conclude, in particular, that if $f: A \to B$ is a function, then A is the domain of f and B contains the range of f. We call B the *codomain* of $f: A \to B$.

2.4 Corollary Let $f: A \to B$ be a function; if C is any class such that $\operatorname{ran} f \subseteq C$, then $f: A \to C$ is a function.

Proof. If $f: A \to B$ is a function, then by 2.3, F2 holds and $\operatorname{dom} f = A$; thus, if $\operatorname{ran} f \subseteq C$, then, by 2.3, $f: A \to C$ is a function. ■

Let $f: A \to B$ be a function and let $x \in A$; it is customary to use the symbol $f(x)$ to designate the image of x. Thus,

$$y = f(x) \quad \textbf{has the same meaning as } (x, y) \in f.$$

When we write $y = f(x)$ instead of $(x, y) \in f$, Conditions F1 and F2 take the form

F1. $\forall x \in A, \exists y \in B, y = f(x)$.

F2. If $y_1 = f(x)$ and $y_2 = f(x)$, then $y_1 = y_2$.

It is often convenient to write F2 in a slightly different way. F2 states that if $(x, y_1) \in f$ and $(x, y_2) \in f$, that is, if

$$x \begin{array}{l} \overset{f}{\nearrow} y_1 \\ \underset{f}{\searrow} y_2 \end{array}$$

then $y_1 = y_2$. This is the same as saying that if $x_1 \overset{f}{\mapsto} f(x_1)$, $x_2 \overset{f}{\mapsto} f(x_2)$, and $x_1 = x_2$; that is, if

$$x_1 = x_2 \begin{array}{l} \overset{f}{\nearrow} f(x_1) \\ \underset{f}{\searrow} f(x_2) \end{array}$$

then $f(x_1) = f(x_2)$. Thus, F2 may be written in the form

F2°. If $x_1 = x_2$, then $f(x_1) = f(x_2)$.

2.5 Theorem Let $f: A \to B$ and $g: A \to B$ be functions. Then $f = g$ if and only if $f(x) = g(x)$, $\forall x \in A$.

Proof. First, let us assume that $f = g$. Then, for arbitrary $x \in A$,

$$y = f(x) \Leftrightarrow (x, y) \in f \Leftrightarrow (x, y) \in g \Leftrightarrow y = g(x);$$

thus, $f(x) = g(x)$.

Conversely, assume that $f(x) = g(x)$, $\forall x \in A$. Then

$$(x, y) \in f \Leftrightarrow y = f(x) \Leftrightarrow y = g(x) \Leftrightarrow (x, y) \in g;$$

thus, $f = g$. ■

Injective, Surjective and Bijective Functions

The following definitions are of great importance in the study of functions.

2.6 Definition A function $f: A \to B$ is said to be *injective* if it has the following property.

INJ. If $(x_1, y) \in f$ and $(x_2, y) \in f$, then $x_1 = x_2$.

The reader should note that INJ states, simply, that if y is any element of B, then

y has no more than one pre-image;

for if $(x_1, y) \in f$ and $(x_2, y) \in f$, that is, if x_1 and x_2 are both pre-images of y, then INJ dictates that $x_1 = x_2$.

It is often convenient to write INJ in a slightly different way. INJ states that if $(x_1, y) \in f$ and $(x_2, y) \in f$, that is, if

$$\begin{array}{l} x_1 \overset{f}{\searrow} \\ \qquad\quad y \\ x_2 \overset{f}{\nearrow} \end{array}$$

then $x_1 = x_2$. This is the same as saying that if x_1 and x_2 are elements of A and $f(x_1) = f(x_2)$, that is, if

$$\begin{array}{l} x_1 \searrow \\ \qquad f(x_1) = f(x_2), \\ x_2 \nearrow \end{array}$$

then $x_1 = x_2$. Thus, INJ may be written in the form

INJ°. If $f(x_1) = f(x_2)$, then $x_1 = x_2$.

(The function of Fig. 2 is injective, whereas the function of Fig. 3 is not injective.)

2.7 Definition A function $f: A \to B$ is said to be *surjective* if it has the following property:

SURJ. $\forall y \in B, \exists x \in A \ni y = f(x)$.

Clearly, condition SURJ states that *every* element of B is the image of some element of A; that is, $B \subseteq \operatorname{ran} f$. But $\operatorname{ran} f \subseteq B$ by Theorem 2.3; hence **f: $A \to B$ is surjective if and only if $\operatorname{ran} f = B$.** (The function of Fig. 3 is surjective, whereas the function of Fig. 2 is not surjective.)

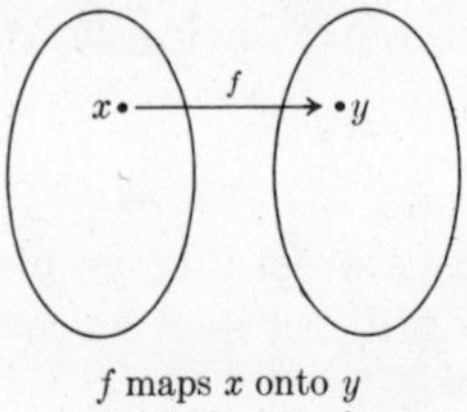

f maps x onto y
y is the image of x

Fig. 1

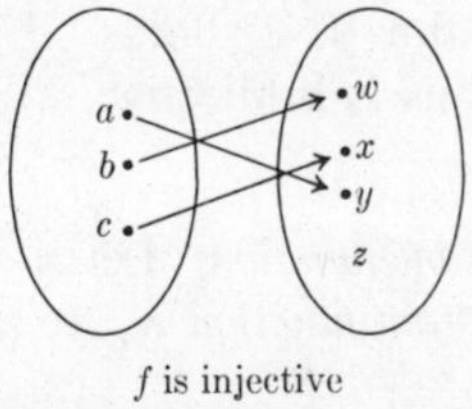

f is injective

Fig. 2

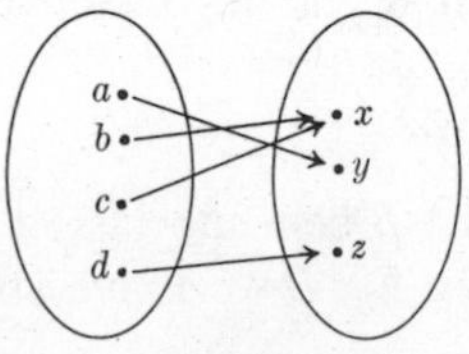

f is surjective

Fig. 3

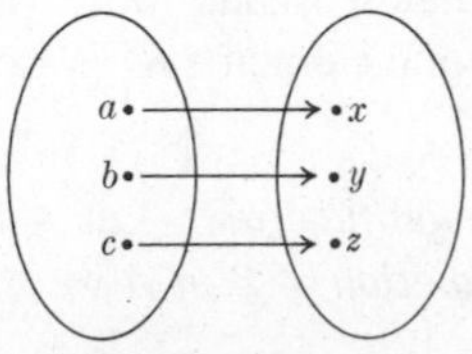

f is bijective

Fig. 4

2.8 Definition A function $f: A \to B$ is said to be *bijective* if it is both injective and surjective.

To say that $f: A \to B$ is injective is to say that every element of B is the image of *no more than one* element of A; to say that f is surjective is to say that every element of B is the image of *at least one* element of A; thus, to say that f is bijective is to say that every element of B is the image of *exactly one* element of A (Fig. 4). In other words, if $f: A \to B$ is a bijective function, every element of A has exactly one image in B and every element of B has exactly one pre-image in A; thus all the elements of A and all the elements of B are associated in pairs; for this reason, if f is bijective, it is sometimes called a *one-to-one correspondence* between A and B.

2.9 Definition If there exists a bijective function $f: A \to B$, then we say that *A and B are in one-to-one correspondence.*

Examples of Functions

2.10 *Identity function.* Let A be a class; by the *identity function on A* we mean the function $I_A: A \to A$ given by

$$I_A(x) = x, \qquad \forall x \in A.$$

In other words,

$$I_A = \{(x, x) \mid x \in A\}.$$

I_A is clearly injective, for suppose $I_A(x) = I_A(y)$; now $I_A(x) = x$ and $I_A(y) = y$, so $x = y$; thus INJ° holds. I_A is surjective because, obviously, the range of I_A is A. Thus I_A is bijective.

2.11 *Constant function.* Let A and B be classes, and let b be an element of B. By the *constant function* K_b we mean the function $K_b: A \to B$ given by

$$K_b(x) = b, \qquad \forall x \in A.$$

In other words, $K_b = \{(x, b) \mid x \in A\}$.

Note that if A has more than one element, K_b is not injective; if B has more than one element, K_b is not surjective.

2.12 *Inclusion function.* Let A be a class and let B be a subclass of A. By the *inclusion function of B in A* we mean the function $E_B: B \to A$ given by

$$E_B(x) = x, \qquad \forall x \in B.$$

Note that if $B = A$, the inclusion function coincides with the identity function

I_A. By the argument used in 2.10, E_B is injective; however, if $B \neq A$, then E_B is not surjective.

2.13 *Characteristic function.* Let 2 designate a class of two elements, say the class $\{0, 1\}$. If A is a class and B is a subclass of A, the *characteristic function of B in A* is the function $C_B: A \to 2$ given by

$$C_B(x) = \begin{matrix} 0 & \text{if} & x \in B, \\ 1 & \text{if} & x \notin B, \end{matrix} \qquad \forall x \in A.$$

Thus C_B maps every element of B onto 0 and every element of $A - B$ onto 1.

2.14 *Restriction of a function.* Let $f: A \to B$ be a function and let C be a subclass of A. By the *restriction of f to C* we mean the function $f_{[C]}: C \to B$ given by

$$f_{[C]}(x) = f(x), \qquad \forall x \in C.$$

To put it another way, $f_{[C]} = \{(x, y) \mid (x, y) \in f \text{ and } x \in C\}$. Note that $f_{[C]} \subseteq f$.

Restrictions of functions have the following properties, which will be useful to us later.

2.15 Theorem If $f: B \cup C \to A$ is a function, then $f = f_{[B]} \cup f_{[C]}$.

The simple proof of this theorem is left as an exercise for the reader.

2.16 Theorem Let $f_1: B \to A$ and $f_2: C \to A$ be functions, where $B \cap C = \varnothing$. If $f = f_1 \cup f_2$, then the following hold:

i) $f: B \cup C \to A$ is a function.
ii) $f_1 = f_{[B]}$ and $f_2 = f_{[C]}$.
iii) If $x \in B$ then $f(x) = f_1(x)$, and if $x \in C$ then $f(x) = f_2(x)$.

Proof. We will begin by proving the following two relations.

a) $(x, y) \in f$ and $x \in B \Leftrightarrow (x, y) \in f_1$.

b) $(x, y) \in f$ and $x \in C \Leftrightarrow (x, y) \in f_2$.

If $(x, y) \in f_1$, then $x \in B$ because $\text{dom } f_1 = B$, and $(x, y) \in f$ because $f = f_1 \cup f_2$. Conversely, suppose $(x, y) \in f$ and $x \in B$: $(x, y) \in f$ implies that $(x, y) \in f_1$ or $(x, y) \in f_2$; if $(x, y) \in f_2$, then $x \in C$ (because $\text{dom } f_2 = C$), which is impossible because $x \in B$ and $B \cap C = \varnothing$; thus, $(x, y) \in f_1$. This proves (**a**); the proof of (**b**) is analogous. Next, we will prove that

c) $\text{dom } f = B \cup C$ and $\text{ran } f \subseteq A$.

Indeed, by 1.43,

$$\operatorname{dom}(f_1 \cup f_2) = \operatorname{dom} f_1 \cup \operatorname{dom} f_2 = B \cup C,$$

and

$$\operatorname{ran}(f_1 \cup f_2) = \operatorname{ran} f_1 \cup \operatorname{ran} f_2 \subseteq A.$$

Our next step will be to prove that

d) f satisfies Condition F2.

Suppose $(x, y_1) \in f$ and $(x, y_2) \in f$; now $x \in \operatorname{dom} f$, so by (**c**), $x \in B$ or $x \in C$. If $x \in B$, then, by (**a**), $(x, y_1) \in f_1$ and $(x, y_2) \in f_1$, so by 2.1, $y_1 = y_2$; if $x \in C$, then, by (**b**), $(x, y_1) \in f_2$ and $(x, y_2) \in f_2$, so by 2.1, $y_1 = y_2$; this proves (**d**). From (**c**), (**d**), and Theorem 2.3, we conclude that $f: B \cup C \to A$ is a function.

By (**a**) and 2.14, $(x, y) \in f_1 \Leftrightarrow (x, y) \in f_{[B]}$, that is, $f_1 = f_{[B]}$; analogously, $f_2 = f_{[C]}$.

Finally, (**a**) states that

$$[y = f(x) \text{ and } x \in B] \Leftrightarrow y = f_1(x)$$

and (**b**) states that

$$[y = f(x) \text{ and } x \in C] \Leftrightarrow y = f_2(x);$$

thus (iii) holds. ■

EXERCISES 2.2

1. Prove that the functions introduced in 2.10 through 2.14 qualify as functions under Definition 2.1.
2. Prove that if $f: A \to B$ is an injective function and $C \subseteq A$, then $f_{[C]}: C \to B$ is an injective function.
3. Let A be a class and let $f = \{(x, (x, x)) \mid x \in A\}$. Show that f is a bijective function from A to I_A.
4. Let $f: A \to B$ and $g: A \to B$ be functions. Prove that if $f \subseteq g$ then $f = g$.
5. Let $f: A \to B$ and $g: C \to D$ be functions. The *product* of f and g is the function defined as follows:

$$[f \cdot g](x, y) = (f(x), g(y)) \quad \text{for every} \quad (x, y) \in A \times C.$$

Prove that $f \cdot g$ is a function from $A \times C$ to $B \times D$. Prove that if f and g are injective, then $f \cdot g$ is injective, and if f and g are surjective, then $f \cdot g$ is surjective. Prove that $\operatorname{ran}[f \cdot g] = (\operatorname{ran} f) \times (\operatorname{ran} g)$.

6. If $f: B \cup C \to A$ is a function, prove that $f = f_{[B]} \cup f_{[C]}$.
7. Let $f_1 : A \to B$ and $f_2: C \to D$ be bijective functions, where $A \cap C = \varnothing$ and $B \cap D = \varnothing$. Let $f = f_1 \cup f_2$; prove that $f: A \cup C \to B \cup D$ is a bijective function.
8. Let $f: B \to A$ and $g: C \to A$ be functions, and suppose that $f_{[B \cap C]} = g_{[B \cap C]}$. If $h = f \cup g$, prove that $h: B \cup C \to A$ is a function, $f = h_{[B]}$ and $g = h_{[C]}$.

9. Let $f: A \to B$ be a function; prove that f is in one-to-one correspondence with A.

By a *functional graph* we mean a graph which satisfies Condition F2. Thus G is a functional graph if and only if

$$(x, y_1) \in G \quad \text{and} \quad (x, y_2) \in G \qquad \text{implies} \qquad y_1 = y_2.$$

10. If G is a functional graph, show that every subclass of G is a functional graph.
11. Let G be a graph. Prove that G is a functional graph if and only if for arbitrary graphs H and J,

$$(H \cap J) \circ G = (H \circ G) \cap (J \circ G).$$

12. Let G be a functional graph. Prove that G is injective if and only if for arbitrary graphs J and H,

$$G \circ (H \cap J) = (G \circ H) \cap (G \circ J).$$

3 PROPERTIES OF COMPOSITE FUNCTIONS AND INVERSE FUNCTIONS

The following theorems express a few basic properties of functions.

2.17 Theorem If $f: A \to B$ and $g: B \to C$ are functions, then $g \circ f: A \to C$ is a function.

Proof

i) By 1.38, dom $g \circ f =$ dom $f = A$; by 1.37(iv), ran $g \circ f \subseteq$ ran $g \subseteq C$.

ii) Suppose $(x, z_1) \in g \circ f$ and $(x, z_2) \in g \circ f$; by 1.34, $\exists y_1 \ni (x, y_1) \in f$ and $(y_1, z_1) \in g$ and $\exists y_2 \ni (x, y_2) \in f$ and $(y_2, z_2) \in g$. From $(x, y_1) \in f$ and $(x, y_2) \in f$ we conclude, by 2.1, that $y_1 = y_2$; thus $(y_1, z_1) \in g$ and $(y_1, z_2) \in g$. It follows by F2 (applied to g) that $z_1 = z_2$; thus, $g \circ f$ satisfies F2.

From (i), (ii), and 2.3, we conclude that $g \circ f: A \to C$ is a function. ■

By 1.34, $(x, y) \in g \circ f$ if and only if for some element z, $(x, z) \in f$ and $(z, y) \in g$. Thus, $x \overset{g \circ f}{\longmapsto} y$ if and only if for some z, $x \overset{f}{\longmapsto} z$ and $z \overset{g}{\longmapsto} y$. (The reader may, if he wishes, picture this statement as in Fig. 5.) This is the same as saying that $y = [g \circ f]$ if and only if for some z, $z = f(x)$ and $y = g(z)$. Thus

2.18 $$[g \circ f](x) = g(f(x)).$$

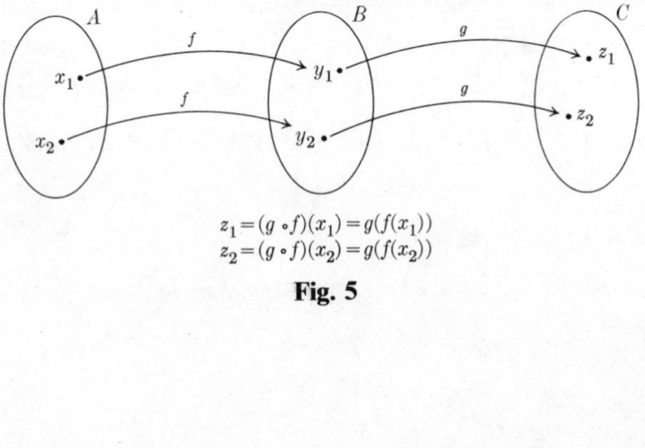

$z_1 = (g \circ f)(x_1) = g(f(x_1))$
$z_2 = (g \circ f)(x_2) = g(f(x_2))$

Fig. 5

2.19 Definition A function $f: A \to B$ is said to be *invertible* if $f^{-1}: B \to A$ is a function.

Let $f: A \to B$ be an invertible function; by 1.33, $(x, y) \in f$ if and only if $(y, x) \in f^{-1}$. Thus $x \overset{f}{\mapsto} y$ if and only if $y \overset{f^{-1}}{\mapsto} x$. (The reader may, if he wishes, picture this statement as in Fig. 6.) Thus

2.20 $y = f(x)$ **if and only if** $x = f^{-1}(y)$.

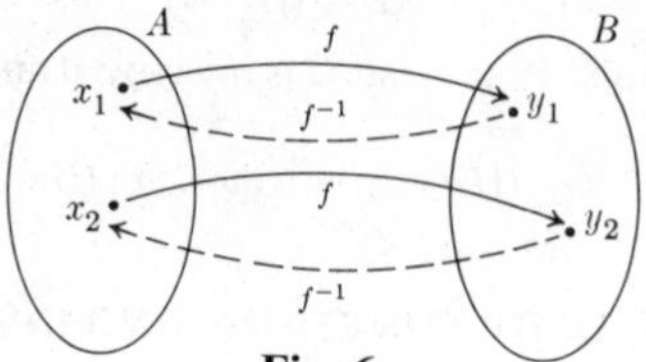

Fig. 6

The next two theorems give a necessary and sufficient condition for a function to be invertible.

2.21 Theorem If $f: A \to B$ is a bijective function, then $f^{-1}: B \to A$ is a bijective function.

Proof. By 2.3, $\operatorname{dom} f = A$, and by 2.7, $\operatorname{ran} f = B$; thus, by 1.37, $\operatorname{dom} f^{-1} = B$ and $\operatorname{ran} f^{-1} = A$. Now we will prove that f^{-1} satisfies F2:

$$\begin{aligned}(y, x_1) \in f^{-1} \text{ and } (y, x_2) \in f^{-1} &\Rightarrow (x_1, y) \in f \text{ and } (x_2, y) \in f && \text{by 1.33}\\ &\Rightarrow x_1 = x_2 && \text{by 2.6.}\end{aligned}$$

Thus, by Theorem 2.3, $f^{-1}: B \to A$ is a function.

Next, we will prove that f^{-1} satisfies INJ:

$$\begin{aligned}(y_1, x) \in f^{-1} \text{ and } (y_2, x) \in f^{-1} &\Rightarrow (x, y_1) \in f \text{ and } (x, y_2) \in f && \text{by 1.33}\\ &\Rightarrow y_1 = y_2 && \text{by 2.1.}\end{aligned}$$

Finally, f^{-1} satisfies SURJ because (see above) $\operatorname{ran} f^{-1} = A$. ■

2.22 Theorem If $f: A \to B$ is invertible, then $f: A \to B$ is bijective.

Proof. Let $f: A \to B$ be invertible; that is, let $f^{-1}: B \to A$ be a function. By 2.3, $\operatorname{dom} f^{-1} = B$, so by 1.37(ii), $\operatorname{ran} f = B$; thus, $f: A \to B$ is surjective. Now

$$\begin{aligned}(x_1, y) \in f \text{ and } (x_2, y) \in f &\Rightarrow (y, x_1) \in f^{-1} \text{ and } (y, x_2) \in f^{-1} && \text{by 1.33}\\ &\Rightarrow x_1 = x_2 && \text{by F2 (applied to } f^{-1}).\end{aligned}$$

Thus $f: A \to B$ is injective. ■

Theorems 2.21 and 2.22 may be summarized as follows:

$f: A \to B$ is invertible if and only if it is bijective; furthermore, if $f: A \to B$ is invertible, then $f^{-1}: B \to A$ is bijective.

The next two theorems give another useful characterization of invertible functions.

2.23 Theorem Let $f: A \to B$ be an invertible function. Then
i) $f^{-1} \circ f = I_A$, and ii) $f \circ f^{-1} = I_B$.

Proof

i) Let $x \in A$ and let $y = f(x)$; then by 2.20, $x = f^{-1}(y)$. Thus

$$[f^{-1} \circ f](x) = f^{-1}[f(x)] = f^{-1}(y) = x = I_A(x);$$

this holds for every $x \in A$, so by 2.5, $f^{-1} \circ f = I_A$.

ii) The proof is analogous to (i), and is left as an exercise. ■

2.24 Theorem Let $f: A \to B$ and $g: B \to A$ be functions. If $g \circ f = I_A$ and $f \circ g = I_B$, then $f: A \to B$ is bijective (hence invertible), and $g = f^{-1}$.

Proof

i) First, we will prove that $f: A \to B$ is injective.

$$f(x_1) = f(x_2) \Rightarrow g(f(x_1)) = g(f(x_2)) \quad \text{by F2° (applied to } g\text{)}$$
$$\Rightarrow [g \circ f](x_1) = [g \circ f](x_2) \quad \text{by 2.18}$$
$$\Rightarrow x_1 = x_2 \quad \text{because } g \circ f = I_A.$$

ii) Next, we will prove that $f: A \to B$ is surjective. If $y \in B$ then $y = I_B(y) = [f \circ g](y) = f(g(y))$; in other words, if y is any element of B, then $y = f(x)$, where $x = g(y) \in A$.

iii) Finally, we will prove that $g = f^{-1}$. To begin with,

$$x = g(y) \Rightarrow f(x) = f(g(y)) = [f \circ g](y) = I_B(y) = y$$
$$\Rightarrow x = f^{-1}(y);$$

conversely,

$$x = f^{-1}(y) \Rightarrow y = f(x)$$
$$\Rightarrow g(y) = g(f(x)) = [g \circ f](x) = I_A(x) = x.$$

Thus, $\forall y \in B$, $x = f^{-1}(y)$ iff $x = g(y)$; that is, $f^{-1}(y) = g(y)$; it follows (by 2.5) that $f^{-1} = g$. ■

Theorems 2.23 and 2.24 may be summarized as follows:

$f: A \to B$ is invertible if and only if there exists a function $g: B \to A$ such that

$g \circ f = I_A$ and $f \circ g = I_B$. The function g, if it exists, is the inverse of f.

Our next theorem gives an important characterization of injective functions.

2.25 Theorem Let $f: A \to B$ be a function; $f: A \to B$ is injective if and only if there exists a function $g: B \to A$ such that $g \circ f = I_A$.

Proof

i) Suppose there exists a function $g: B \to A$ such that $g \circ f = I_A$. To prove that $f: A \to B$ is injective, we repeat part (i) of the proof of 2.24.

ii) Conversely, suppose that $f: A \to B$ is injective; let $C = \operatorname{ran} f$. By 2.4, $f: A \to C$ is a function; $f: A \to C$ is surjective (because $C = \operatorname{ran} f$), hence it is bijective; thus $f^{-1}: C \to A$ is a function. If a is some fixed element of A, let $K_a:(B - C) \to A$ be the constant function (see 2.11) which maps every element of $B - C$ onto a. If $g = f^{-1} \cup K_a$, then, by 2.16(i), $g: B \to A$ is a function (see Fig. 7). Finally, if $x \in A$, let $y = f(x)$; then

$$\begin{aligned} [g \circ f](x) = g(f(x)) &= g(y) && \text{because } y = f(x) \\ &= f^{-1}(y) && \text{by 2.16(iii)} \\ &= x && \text{because } x = f^{-1}(y) \text{ by 2.20.} \end{aligned}$$

Thus $\forall x \in A$, $[g \circ f](x) = I_A(x)$; it follows by 2.5 that $g \circ f = I_A$. ■

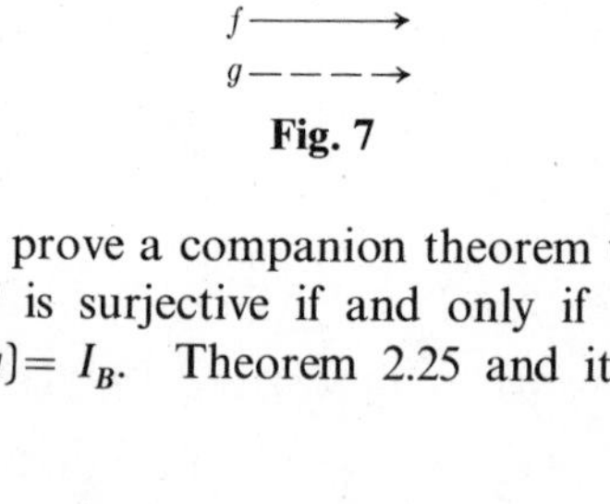

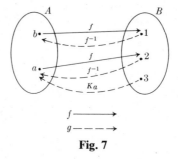

Fig. 7

In Chapter 5 we will prove a companion theorem to 2.25, which will state the following: $f: A \to B$ is surjective if and only if there exists a function $g: B \to A$ such that $[f \circ g] = I_B$. Theorem 2.25 and its companion are often paraphrased as follows.

Let $f: A \to B$ be a function; $f: A \to B$ is injective if and only if it has a "left inverse" and surjective if and only if it has a "right inverse".

2.26 Theorem Suppose $f: A \to B$, $g: B \to C$, and $g \circ f: A \to C$ are functions.

i) If f and g are injective, then $g \circ f$ is injective.
ii) If f and g are surjective, then $g \circ f$ is surjective.
iii) If f and g are bijective, then $g \circ f$ is bijective.

Proof

i) Suppose that f and g both satisfy INJ°: then

$$g[f(x_1)] = g[f(x_2)] \Rightarrow f(x_1) = f(x_2) \Rightarrow x_1 = x_2;$$

thus $g \circ f$ satisfies INJ°.

ii) Suppose that f and g both satisfy SURJ: if $z \in C$, then $\exists y \in B \ni x = g(y)$; since $y \in B$, $\exists x \in A \ni y = f(x)$; thus $z = g(f(x)) = [g \circ f](x)$. Consequently, $g \circ f$ satisfies SURJ.

iii) This follows immediately from (i) and (ii). ■

It follows from 2.26(iii) that

the composite of two invertible functions is invertible.

Furthermore, by 1.35(iii), $(g \circ f)^{-1} = f^{-1} \circ g^{-1}$.

EXERCISES 2.3

1. Let $f: A \to B$ be a function. Prove that $I_B \circ f = f$ and $f \circ I_A = f$.
2. Suppose $f: A \to B$ and $g: B \to C$ are functions. Prove that if $g \circ f$ is injective, then f is injective; prove that if $g \circ f$ is surjective, then g is surjective. Conclude that if $g \circ f$ is bijective, then f is injective and g is surjective.
3. Give an example to show that the converse of the last statement of Exercise 2 does not hold.
4. Let $f: A \to B$ and $g: B \to A$ be functions. Suppose that $y = f(x)$ if and only if $x = g(y)$. Prove that f is invertible and $g = f^{-1}$.
5. Let $g: B \to C$ and $h: B \to C$ be functions. Suppose that $g \circ f = h \circ f$ for every function $f: A \to B$. Prove that $g = h$.
6. Suppose $g: A \to B$ and $h: A \to B$ are functions. Let C be a set with more than one element; suppose that $f \circ g = f \circ h$ for every function $f: B \to C$. Prove that $g = h$.
7. Let $f: B \to C$ be a function. Prove that f is injective if and only if, for every pair of functions $g: A \to B$ and $h: A \to B, f \circ g = f \circ h \Rightarrow g = h$.
8. Let $f: A \to B$ be a function. Prove that f is surjective if and only if, for every pair of functions $g: B \to C$ and $h: B \to C, g \circ f = h \circ f \Rightarrow g = h$.
9. Let $f: A \to C$ and $g: A \to B$ be functions. Prove that there exists a function $h: B \to C$ such that $f = h \circ g$ if and only if $\forall x, y \in A$,

$$g(x) = g(y) \Rightarrow f(x) = f(y).$$

Prove that h is unique.

10. Let $f: C \to A$ and $g: B \to A$ be functions, and suppose that g is bijective. Prove that there exists $h: C \to B$ such that $f = g \circ h$ if and only if $\operatorname{ran} f \subseteq \operatorname{ran} g$. Prove that h is unique.

11. Let $f: A \to B$ be a function, and let $C \subseteq A$. Prove that $f_{[C]} = f \circ E_C$, where E_C is the inclusion function of C in A (2.12).

4 DIRECT IMAGES AND INVERSE IMAGES UNDER FUNCTIONS

2.27 Definition Let $f: A \to B$ be a function; if C is any subclass of A, the direct image of C under f, which we write $\overline{f}(C)$, is the following subclass of B:

$$\overline{f}(C) = \{y \in B \mid \exists x \in C \ni y = f(x)\}.$$

That is, $\overline{f}(C)$ is the class of all the images of elements in C.

2.28 Definition Let $f: A \to B$ be a function; if D is any subclass of B, the inverse image of D under f, which we write $\overline{\overline{f}}(D)$, is the following subclass of A:

$$\overline{\overline{f}}(D) = \{x \in A \mid f(x) \in D\}.$$

That is, $\overline{\overline{f}}(D)$ is the class of all the pre-images of elements in D.

If $\{a\}$ and $\{b\}$ are singletons, we will write $\overline{f}(a)$ for $\overline{f}(\{a\})$ and $\overline{f}(b)$ for $\overline{f}(\{b\})$.

2.29 Theorem Let $f: A \to B$ be a function.

i) If $C \subseteq A$ and $D \subseteq A$, then $C = D \Rightarrow \overline{f}(C) = \overline{f}(D)$.

ii) If $C \subseteq B$ and $D \subseteq B$, then $C = D \Rightarrow \overline{\overline{f}}(C) = \overline{\overline{f}}(D)$.

Proof

i) Suppose $C = D$; then

$$\begin{aligned} y \in \overline{f}(C) &\Leftrightarrow \exists x \in C \ni y = f(x) && \text{by 2.27} \\ &\Leftrightarrow \exists x \in D \ni y = f(x) && \text{because } C = D \\ &\Leftrightarrow y \in \overline{f}(D) && \text{by 2.27.} \end{aligned}$$

ii) Suppose $C = D$; then

$$\begin{aligned} x \in \overline{\overline{f}}(C) &\Leftrightarrow f(x) \in C && \text{by 2.28} \\ &\Leftrightarrow f(x) \in D && \text{because } C = D \\ &\Leftrightarrow x \in \overline{\overline{f}}(D) && \text{by 2.28.} \ \blacksquare \end{aligned}$$

Caution. $\overline{f}(C) = \overline{f}(D)$ does not always imply that $C = D$; for a simple counter-

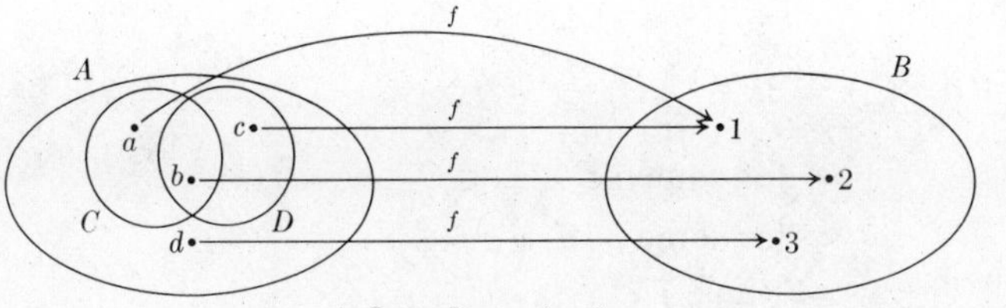

$\overline{f}(C) = \overline{f}(D) = \{1, 2\}$.

Fig. 8

example, see Fig. 8. Similarly, $\overline{\overline{f}}(C) = \overline{\overline{f}}(D)$ does not always imply $C = D$; for a counterexample, the reader should look at Fig. 9. (*However, see Exercise* 3, *Exercise Set* 2.4.)

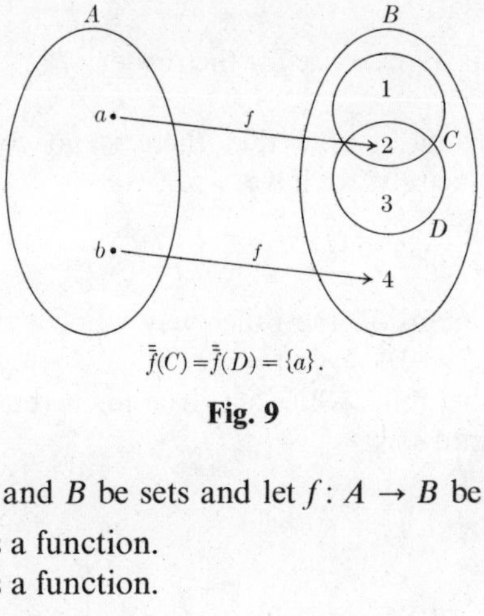

$\overline{\overline{f}}(C) = \overline{\overline{f}}(D) = \{a\}$.

Fig. 9

2.30 Theorem Let A and B be sets and let $f: A \to B$ be a function; then

i) $\overline{f}: \mathscr{P}(A) \to \mathscr{P}(B)$ is a function.
ii) $\overline{\overline{f}}: \mathscr{P}(B) \to \mathscr{P}(A)$ is a function.

Proof

i) By 2.27, it is easy to see that $\operatorname{dom} \overline{f} = \mathscr{P}(A)$ and $\operatorname{ran} \overline{f} \subseteq \mathscr{P}(B)$. Theorem 2.29(i) states that $\overline{f}$ satisfies Condition F2; thus by 2.3, $\overline{f}: \mathscr{P}(A) \to \mathscr{P}(B)$ is a function.
ii) Analogously, $\overline{\overline{f}}: \mathscr{P}(B) \to \mathscr{P}(A)$ is a function. ■

2.31 Theorem Let $f: A \to B$ be a function, let $\{C_i\}_{i \in I}$ be a family of subclasses of A, and let $\{D_i\}_{i \in I}$ be a family of subclasses of B. Then

i) $\overline{f}(\bigcup_{i \in I} C_i) = \bigcup_{i \in I} \overline{f}(C_i)$,
ii) $\overline{\overline{f}}(\bigcup_{i \in I} D_i) = \bigcup_{i \in I} \overline{\overline{f}}(D_i)$,
iii) $\overline{\overline{f}}(\bigcap_{i \in I} D_i) = \bigcap_{i \in I} \overline{\overline{f}}(D_i)$.

Proof

i) $y \in \overline{f}(\bigcup_{i \in I} C_i) \Leftrightarrow \exists x \in \bigcup_{i \in I} C_i \ni y = f(x)$ by 2.27

$\Leftrightarrow$ for some $j \in I$, $\exists x \in C_j \ni y = f(x)$ by 1.39

$\Leftrightarrow$ for some $j \in I$, $y \in \overline{f}(C_j)$ by 2.27

$\Leftrightarrow y \in \bigcup_{i \in I} \overline{f}(C_i)$ by 1.39.

ii) $x \in \overline{\overline{f}}(\bigcup_{i \in I} D_i) \Leftrightarrow f(x) \in \bigcup_{i \in I} D_i$ by 2.28

$\Leftrightarrow$ for some $j \in I$, $f(x) \in D_j$ by 1.39

$\Leftrightarrow$ for some $j \in I$, $x \in \overline{\overline{f}}(D_j)$ by 2.28

$\Leftrightarrow x \in \bigcup_{i \in I} \overline{\overline{f}}(D_i)$ by 1.39.

iii) The proof is left as an exercise for the reader. ∎

Caution. It is important to note that there is no counterpart of Theorem 2.31(iii) for $\overline{f}$; more precisely, we have

$$\overline{f}(\bigcap_{i \in I} C_i) \subseteq \bigcap_{i \in I} \overline{f}(C_i),$$

but we do not have inclusion the other way. For a simple counterexample, see Fig. 8, where $\overline{f}(C \cap D) = \overline{f}(b) = \{2\} \neq \{1, 2\} = \overline{f}(C) \cap \overline{f}(D)$. For this reason, a variety of theorems which are true for inverse images of sets fail to hold for direct images of sets.

EXERCISES 2.4

1. Suppose that $f: A \to B$ is a function, $C \subseteq A$ and $D \subseteq B$.
   a) Prove that $C \subseteq \overline{\overline{f}}[\overline{f}(C)]$. b) Prove that $\overline{f}[\overline{\overline{f}}(D)] \subseteq D$.
2. Suppose that $f: A \to B$ is a function, $C \subseteq A$ and $D \subseteq B$.
   a) If f is injective, prove that $C = \overline{\overline{f}}[\overline{f}(C)]$.
   b) If f is surjective, prove that $D = \overline{f}[\overline{\overline{f}}(D)]$.
3. Let $f: A \to B$ be a function. Prove the following.
   a) Suppose $C \subseteq A$ and $D \subseteq A$; if f is injective, then $\overline{f}(C) = \overline{f}(D) \Rightarrow C = D$.
   b) Suppose $C \subseteq B$ and $D \subseteq B$; if f is surjective, then $\overline{\overline{f}}(C) = \overline{\overline{f}}(D) \Rightarrow C = D$.
   [*Hint:* Use the result of Exercise 2.]
4. Let $f: A \to B$ be a function. Prove the following:
   a) If f is injective, then $\overline{\overline{f}} \circ \overline{f}$ is bijective. [*Hint:* Use the result of Exercise 2(a).]
   b) If f is surjective, then $\overline{f} \circ \overline{\overline{f}}$ is bijective. [*Hint:* Use the result of Exercise 2(b).]

5. Suppose that $f: A \to B$ is a function; let $C \subseteq A$.
   a) Prove that $\overline{f}\{\overline{\overline{f}}[\overline{f}(C)]\} = \overline{f}(C)$.
   b) Use the result of (a) to prove that $\overline{f} \circ \overline{\overline{f}} \circ \overline{f} = \overline{f}$.

6. Let $f: A \to B$ be a function. Prove the following:
   a) If f is injective, then $\overline{f}$ is injective.
   b) If f is surjective, then $\overline{\overline{f}}$ is surjective.
   c) If f is bijective, then $\overline{f}$ is bijective.

7. Let $f: A \to B$ be a function. Prove the following:
   a) If f is injective, then $\overline{\overline{f}}$ is surjective.
   b) If f is surjective, then $\overline{\overline{f}}$ is injective.
   c) If f is bijective, then $\overline{\overline{f}}$ is bijective.

8. Let $f: A \to B$ be a function. Prove that

$$\overline{f}(C \cap D) = \overline{f}(C) \cap \overline{f}(D)$$

for every pair of subclasses $C \subseteq A$ and $D \subseteq A$ if and only if f is injective.

9. Suppose that $f: A \to B$ is a function, $C \subseteq B$ and $D \subseteq B$. Prove that

$$\overline{\overline{f}}(C - D) = \overline{\overline{f}}(C) - \overline{\overline{f}}(D).$$

10. Let $f: A \to B$ be a function. Prove each of the following:
    a) If $C \subseteq A$ and $D \subseteq A$, then $\overline{f}(C) - \overline{f}(D) \subseteq \overline{f}(C - D)$.
    b) $\overline{f}(C) - \overline{f}(D) = \overline{f}(C - D)$ for every pair of sublcasses $C \subseteq A$ and $D \subseteq A$ if and only if f is injective.

5 PRODUCT OF A FAMILY OF SETS

In this section we generalize the notion of Cartesian product, introduced in Chapter 1, Section 4, and define the Cartesian product of an arbitrary family of sets. Note that we confine our attention, here, to *sets*; we need to do this because—as the reader will recall—we may form arbitrary classes of *sets*, but not arbitrary classes of *classes*.

The product of two sets A and B has been defined to be the set $A \times B$ of all ordered pairs (x, y), where $x \in A$ and $y \in B$. This definition may be extended, in a natural way, to a finite number of sets $A_1, A_2, \ldots, A_n$; we may define the product $A_1 \times A_2 \times \cdots \times A_n$ to be the set of all "ordered n-tuples" $(a_1, a_2, \ldots, a_n)$, where $a_i \in A_i$ for each index $i = 1, 2, \ldots, n$. Now, we wish to extend this concept to the case of an indexed family of sets, $\{A_i\}_{i \in I}$, where the index set I is any set whatsoever. Evidently we cannot speak of "I-tuples" of elements, because I may be an infinite set and may not be ordered; therefore, we must alter our approach to the problem.

Let us take another look at the product $A_1 \times A_2 \times \cdots \times A_n$. Clearly $\{A_1, A_2, \ldots, A_n\}$ is a family whose index set is $I = \{1, 2, \ldots, n\}$. Now, *an ordered n-tuple may be regarded as a function* (*whose domain is I*) *which maps* *each element $i \in I$ onto an element a_i in A_i.* Indeed, if f is such a function, then f is described by the following table.

x	$f(x)$
1	a_1
2	a_2
.	
.	
.	
n	a_n

Using the table, we may construct the ordered n-tuple $(a_1, a_2, \ldots, a_n)$; conversely, if we are given the ordered n-tuple $(a_1, a_2, \ldots, a_n)$, then we may construct the table; (in fact, the ordered n-tuple is simply the table presented as a horizontal array). Thus, the function f and the ordered n-tuple $(a_1, a_2, \ldots, a_n)$ are, essentially, one and the same thing. This simple observation leads to the following definition of the product of a family of sets.

2.32 Definition Let $\{A_i\}_{i \in I}$ be a family of sets, where I is a set; let

$$A = \bigcup_{i \in I} A_i.$$

The *product* of the sets A_i is defined to be the class

$$\prod_{i \in I} A_i = \{f \mid f\colon I \to A \text{ is a function, and } f(i) \in A_i, \forall i \in I\}.$$

The legitimacy of this definition is established in the next section.

2.33 Example Let $I = \{1, 2\}$, $A_1 = \{a, b\}$, and $A_2 = \{c, d\}$. By 2.32, $\prod_{i \in I} A_i$ consists of all the functions $f\colon \{1, 2\} \to \{a, b, c, d\}$ such that $f(1) \in A_1$ and $f(2) \in A_2$. There are four such functions, given by the following tables.

x	$f(x)$
1	a
2	c

x	$f(x)$
1	a
2	d

x	$f(x)$
1	b
2	c

x	$f(x)$
1	b
2	d

We may identify these four functions with the four ordered pairs (a, c), (a, d), (b, c), and (b, d), respectively. Thus $\prod_{i \in I} A_i$ is exactly $A_1 \times A_2$.

We adopt the following notational convention: henceforth, we will designate elements of a product $\prod_{i \in I} A_i$ by bold face letters **a**, **b**, **c**, etc.

If **a** is an element of $\prod_{i \in I} A_i$ and $j \in I$, we agree that $\mathbf{a}_j$ will have the same meaning as $\mathbf{a}(j)$; we will call $\mathbf{a}_j$ the *j-coordinate* of **a**.

Let $\{A_i\}_{i \in I}$ be an indexed family, and, for each $i \in I$, let $x_i \in A_i$. We will use the symbol $(x_i)_{i \in I}$ to designate the element in $\prod_{i \in I} A_i$ whose i-coordinate, for each $i \in I$, is x_i.

Let $A = \prod_{i \in I} A_i$; corresponding to each index $i \in I$, we define a function p_i from A to A_i by

$$p_i(\mathbf{a}) = \mathbf{a}_i, \quad \forall \mathbf{a} \in A.$$

The function p_i is called the *i-projection* of A to A_i.

2.34 Definition If A and B are arbitrary sets, the symbol B^A refers to the class of all functions from A to B.

In particular, if 2 denotes a set of two elements, then 2^A denotes the class of all functions from A to 2. The following is an important result which will be used in a later chapter.

2.35 Theorem If A is a set, then 2^A and $\mathscr{P}(A)$ are in one-to-one correspondence.

Proof. We will show that there exists a bijective function $\gamma: \mathscr{P}(A) \to 2^A$. If $B \in \mathscr{P}(A)$, let C_B denote the characteristic function of B in A (see 2.13); C_B is an element of 2^A. We define γ by

$$\gamma(B) = C_B, \quad \forall B \in \mathscr{P}(A).$$

By the way γ is defined, it is clear that γ maps every $B \in \mathscr{P}(A)$ onto a uniquely determined element of 2^A; hence $\gamma: \mathscr{P}(A) \to 2^A$ is a function; it remains to show that γ is injective and surjective.

i) Let $B, D \in \mathscr{P}(A)$; if $\gamma(B) = \gamma(D)$, then $C_B = C_D$; hence

$$\{x \in A \mid C_B(x) = 0\} = \{x \in A \mid C_D(x) = 0\};$$

that is, $B = D$. Thus γ satisfies INJ°.

ii) If $f \in 2^A$, and if we let $B = \overline{f}(0)$, then $f = C_B = \gamma(B)$. Thus γ satisfies condition SURJ. ■

It is easy to show that if A and B are sets, then A^B is a set (see Exercise 12, Exercise Set 2.5). Using this fact, it can easily be shown that if $\{A_i\}_{i \in I}$ is an index family of *sets* such that the index class I is a *set*, then $\prod_{i \in I} A_i$ is a set (see Exercise 13, Exercise Set 2.5, and Remark 2.37).

EXERCISES 2.5

1. Let $A = \{1, 2, 3\}$, $B = \{a, b\}$. Find A^B, B^A, 2^A, and $\mathscr{P}(A)$.

2. Suppose that $\{B_i\}_{i\in I}$ is a family of subsets of A. Prove that
$$\prod_{i\in I} B_i \subseteq A^I.$$

3. Suppose that $\{A_i\}_{i\in I}$ and $\{B_i\}_{i\in I}$ are families of sets with the same index set I. Show that if $A_i \subseteq B_i$, $\forall i \in I$, then
$$\prod_{i\in I} A_i \subseteq \prod_{i\in I} B_i.$$

4. Suppose that $\{A_i\}_{i\in I}$ and $\{B_i\}_{i\in I}$ are families of nonempty sets with the same index set I. Prove that if
$$\prod_{i\in I} A_i \subseteq \prod_{i\in I} B_i,$$
then $A_i \subseteq B_i$ for each index i.

5. Suppose that $\{A_i\}_{i\in I}$ and $\{B_i\}_{i\in I}$ are families of sets with the same index set I. Prove that
$$(\prod_{i\in I} A_i) \cap (\prod_{i\in I} B_i) = \prod_{i\in I} (A_i \cap B_i).$$

6. Let $\{A_i\}_{i\in I}$ and $\{B_j\}_{j\in J}$ be families of sets. Prove the following:

a) $(\prod_{i\in I} A_i) \cap (\prod_{j\in J} B_j) = \prod_{(i,j)\in I\times J} (A_i \cap B_j)$.

b) $(\prod_{i\in I} A_i) \cup (\prod_{j\in J} B_j) = \prod_{(i,j)\in I\times J} (A_i \cup B_j)$.

7. Let $\{A_i\}_{i\in I}$ be a family of sets, and for each $i \in I$, let B_i be a subset of A_i. Prove that
$$\bigcap_{i\in I} \overline{p}_i(B_i) = \prod_{i\in I} B_i.$$

8. Let $\{A_i\}_{i\in I}$ be an indexed family, and let
$$A = \prod_{i\in I} A_i.$$
If $B \subseteq A$, let $B_i = \overline{p}_i(B)$ for each $i \in I$. Prove that $B \subseteq \prod_{i\in I} B_i$.

9. Prove that $A^C \cup B^C \subseteq (A \cup B)^C$.

10. Prove that $(A \cap B)^C = A^C \cap B^C$.

11. Prove that $(A - B)^C = A^C - B^C$.

12. Prove that if A and B are sets, then A^B is a set. [*Hint:* Each element of A^B is a subset of $B \times A$. Use Axioms A3 through A6.]

13. Let $\{A_i\}_{i\in I}$ be an indexed family; suppose that I is a set, that each A_i is a set, and that $\{A_i \mid i \in I\}$ is a set. Prove that $\prod_{i\in I} A_i$ is a set. [*Hint:* Use the results of Exercises 2 and 12.]

6 THE AXIOM OF REPLACEMENT

Axioms A3 through A6 are "set" axioms, that is, they are designed for the purpose of establishing the properties of sets. We are now in a position to introduce our last "set" axiom. This axiom is motivated by the following considerations.

We noted earlier that we are to think of a *set* as a class which is "not too large." Now, if A and B are classes and $f: A \to B$ is a surjective function, then, in an obvious intuitive sense, B has "as many, or fewer elements than A" (see Fig. 3). Thus if A is "not too large" and $f: A \to B$ is a surjective function, it stands to reason that B is "not too large." These remarks lead us to state the following as an axiom.

A7 If A is a set and $f: A \to B$ is a surjective function, then B is a set.

Statement A7 is traditionally called the *axiom of replacement*; it has the following consequences.

2.36 If A is a set and A is in one-to-one correspondence with B, then B is a set.

2.37 *Remark.* Let $\{A_i\}_{i \in I}$ be an indexed family of sets, where the index class I is a set. It is clear that the function ϕ defined by $\phi(i) = A_i$ is a surjective function from I to $\{A_i \mid i \in I\}$; thus we have

If $\{A_i\}_{i \in I}$ is an indexed family of sets and I is a set, then $\{A_i \mid i \in I\}$ is a set.

Now if $\{A_i\}_{i \in I}$ is a family of sets, where I is a set, it follows from the preceding remark and from Axiom A5 that $\bigcup_{i \in I} A_i$ is a set. Now if $f: I \to A$ is a function where I and A are sets, then $f \subseteq I \times A$, so by Theorem 1.53 and Axiom A4, f is a set. This shows that Definition 2.32 is legitimate.

EXERCISE

Show that Axiom A3 follows from Axiom A7. Thus, Axiom A3 can now be eliminated.

3

Relations

1 INTRODUCTION

Intuitively, a binary relation in a class A is a statement $R(x, y)$ which is either true or false for each ordered pair (x, y) of elements of A. For instance, the relation "*x divides y*," which we may write $D(x, y)$, is a relation in the class $\mathbb{Z}$ of the integers: $D(x, y)$ is true for every pair (x, y) of integers such that y is a multiple of x; it is false for every other pair of integers.

The *representing graph* of a relation in A is a graph $G \subseteq A \times A$ which consists of all the pairs (x, y) such that $R(x, y)$ is true. Conversely, if we are given an arbitrary graph $G \subseteq A \times A$, then G defines a relation in A, namely the relation R such that $R(x, y)$ is true if and only if $(x, y) \in G$.

Thus, as we did in the case of functions, we are able to identify relations with their representing graphs. In this way the study of relations is part of elementary set theory.

2 FUNDAMENTAL CONCEPTS AND DEFINITIONS

3.1 Definition Let A be a class; by a *relation in A* we mean an arbitrary subclass of $A \times A$.

3.2 Definition Let G be a relation in A; then

G is called *reflexive* if

$$\forall x \in A, (x, x) \in G.$$

G is called *symmetric* if

$$(x, y) \in G \Rightarrow (y, x) \in G.$$

G is called *anti-symmetric* if

$$(x, y) \in G \text{ and } (y, x) \in G \Rightarrow x = y.$$

G is called *transitive* if

$$(x, y) \in G \text{ and } (y, z) \in G \Rightarrow (x, z) \in G.$$

3.3 Definition The *diagonal graph* I_A is defined to be the class $\{(x, x) \mid x \in A\}$.

It is easy to see that G is reflexive if and only if $I_A \subseteq G$.

There is a variety of interesting and useful alternative ways of defining the above notions. Some are given in the next theorem.

3.4 Theorem Let G be a relation in A.

i) G is symmetric if and only if $G = G^{-1}$.
ii) G is antisymmetric if and only if $G \cap G^{-1} \subseteq I_A$.
iii) G is transitive if and only if $G \circ G \subseteq G$.

Proof

i) Suppose G is symmetric. Then

$$(x, y) \in G \Leftrightarrow (y, x) \in G \Leftrightarrow (x, y) \in G^{-1};$$

thus $G = G^{-1}$. Conversely, suppose $G = G^{-1}$. Then

$$(x, y) \in G \Rightarrow (x, y) \in G^{-1} \Rightarrow (y, x) \in G.$$

ii) Suppose G is antisymmetric. Then

$$\begin{aligned}(x, y) \in G \cap G^{-1} &\Rightarrow (x, y) \in G \text{ and } (x, y) \in G^{-1}\\ &\Rightarrow (x, y) \in G \text{ and } (y, x) \in G\\ &\Rightarrow x = y\\ &\Rightarrow (x, y) = (x, y) \in I_A.\end{aligned}$$

Conversely, suppose that $G \cap G^{-1} \subseteq I_A$. Then

$$\begin{aligned}(x, y) \in G \text{ and } (y, x) \in G &\Rightarrow (x, y) \in G \text{ and } (x, y) \in G^{-1}\\ &\Rightarrow (x, y) \in G \cap G^{-1} \subseteq I_A\\ &\Rightarrow x = y.\end{aligned}$$

iii) Suppose G is transitive. Then

$(x, y) \in G \circ G \Rightarrow \exists z \ni (x, z) \in G$ and $(z, y) \in G \Rightarrow (x, y) \in G$. Thus $G \circ G \subseteq G$.

Conversely, suppose $G \circ G \subseteq G$: Then $(x, y) \in G$ and $(y, z) \in G \Rightarrow (x, z) \in G \circ G \subseteq G$. ■

3.5 Definition A relation is called an *equivalence relation* if it is reflexive, symmetric,and transitive.

A relation is called an *order relation* if it is reflexive, antisymmetric, and transitive.

3.6 Definition Let G be a relation in A.

G is called *irreflexive* if

$$\forall x \in A, \quad (x, x) \notin G.$$

G is called *asymmetric* if

$$(x, y) \in G \Rightarrow (y, x) \notin G.$$

G is called *intransitive* if

$$(x, y) \in G \text{ and } (y, z) \in G \Rightarrow (x, z) \notin G.$$

Examples Let $\mathbb{Z}$ designate the set of the integers; the equality relation in $\mathbb{Z}$ is reflexive, symmetric, and transitive; hence, it is an equivalence relation. The relation $\leqslant$ ("less than or equal to") is reflexive, antisymmetric, and transitive; hence it is an order relation. The relation $<$ ("strictly less than") is not an order relation: it is irreflexive, asymmetric, and transitive; such a relation is called a relation of *strict order*.

EXERCISES 3.2

1. Each of the following describes a relation in the set $\mathbb{Z}$ of the integers. State, for each one, whether it has any of the following properties: reflexive, symmetric, antisymmetric, transitive, irreflexive, asymmetric, intransitive. Determine whether it is an equivalence relation, an order relation, or neither. Prove your answer in each case.
   a) $G = \{(x, y) \mid x + y < 3\}$.
   b) $G = \{(x, y) \mid x \text{ divides } y\}$.
   c) $G = \{(x, y) \mid x \text{ and } y \text{ are relatively prime}\}$.
   d) $G = \{(x, y) \mid x + y \text{ is an even number}\}$.
   e) $G = \{(x, y) \mid x = y \text{ or } x = -y\}$.
   f) $G = \{(x, y) \mid x + y \text{ is even and } x \text{ is a multiple of } y\}$.
   g) $G = \{(x, y) \mid y = x + 1\}$.

2. Let G be a relation in A; prove each of the following:
   a) G is irreflexive if and only if $G \cap I = \varnothing$.
   b) G is asymmetric if and only if $G \cap G^{-1} = \varnothing$.
   c) G is intransitive if and only if $(G \circ G) \cap G = \varnothing$.

3. Show that if G is an equivalence relation in A, then $G \circ G = G$.

4. Let $\{G_i\}_{i \in I}$ be an indexed family of equivalence relations in A. Show that $\bigcap_{i \in I} G_i$ is an equivalence relation in A.

5. Let $\{G_i\}_{i \in I}$ be an indexed family of order relations in A. Show that $\bigcap_{i \in I} G_i$ is an order relation in A.

6. Let H be a reflexive relation in A. Prove that for any relation G in A, $G \subseteq H \circ G$ and $G \subseteq G \circ H$.

7. Let G and H be relations in A; suppose that G is reflexive and H is reflexive and transitive. Show that $G \subseteq H$ if and only if $G \circ H = H$. (In particular, this holds if G and H are equivalence relations.)

8. Show that the inverse of an order relation in A is an order relation in A.

9. Let G be a relation in A. Show that G is an order relation if and only if $G \cap G^{-1} = I_A$ and $G \circ G = G$.

10. Let G and H be equivalence relations in A. Show that $G \circ H$ is an equivalence relation in A if and only if $G \circ H = H \circ G$.

11. Let G and H be equivalence relations in A. Prove that $G \cup H$ is an equivalence relation in A if and only if $G \circ H \subseteq G \cup H$ and $H \circ G \subseteq G \cup H$.

12. Let G be an equivalence relation in A. Prove that if H and J are arbitrary relations in A, then $G \subseteq H$ and $G \subseteq J \Rightarrow G \subseteq H \circ J$.

3 EQUIVALENCE RELATIONS AND PARTITIONS

In the remainder of this chapter we will concern ourselves with equivalence relations in *sets*. The concepts we are about to introduce arise naturally in terms of sets, but cannot be extended to proper classes; to understand why not, the reader should review our discussion in Section 7 of Chapter 1. Briefly, if A is a set and $P(X)$ is a property, then by 1.52 it is legitimate to form the set of all the subsets $X \subseteq A$ which satisfy $P(X)$. However, if A were an arbitrary class, it would not be permissible to form the "class of all subclasses of A which satisfy $P(X)$." This restriction compels us to confine the following discussion to sets. Intuitively, this should not disturb the reader too much, for a set is almost the same thing as a class: a set is any class except an "excessively large" one.

3.7 Definition Let A be a set; by a *partition* of A we mean a family $\{A_i\}_{i\in I}$ of nonempty subsets of A with the following properties:

P1. $\forall i, j \in I, \quad A_i \cap A_j = \varnothing \qquad \text{or} \qquad A_i = A_j.$

P2. $A = \bigcup_{i\in I} A_i.$

Intuitively, a partition is a famiily of subsets of A which are disjoint from one another, and whose union is all of A (Fig. 1). The subsets are called the *members* of the partition. It is customary to allow a given member of the partition to be designated by more than one index; that is, we may have

$A_i = A_j$, where $i \neq j$. Hence the condition that two *distinct* members be disjoint is correctly expressed by P1.

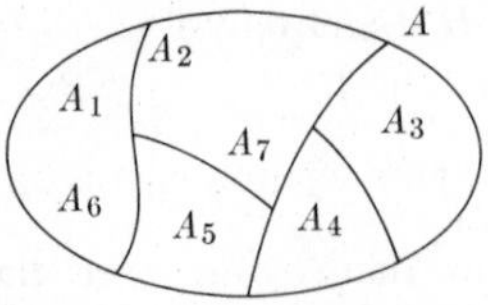

$\{A_1, A_2, A_3, A_4, A_5, A_6, A_7\}$ is a partition of A. Note that $A_1 = A_6$ and $A_2 = A_7$.

Fig. 1

Property P1 states that any two members A_i and A_j are either disjoint or equal; that is, they have either no elements in common or all their elements in common; in other words, if they have so much as one element in common, they have all their elements in common. Thus, P1 may also be stated as follows:

P1°. If $\exists x \in A_i \cap A_j$, then $A_i = A_j$.

P2 may be replaced by the simpler condition

P2′. $A \subseteq \bigcup_{i \in I} A_i$.

For, independently of Condition P2, we are given that each A_i is a subset of A; hence, by 1.40(i), $\bigcup_{i \in I} A_i \subseteq A$. Consequently, it is sufficient to state P2′ in order to have $A = \bigcup_{i \in I} A_i$. It is convenient to write P2′ in the form

P2°. If $x \in A$, then $x \in A_i$ for some $i \in I$.

Briefly, then, a partition of A is a family $\{A_i\}_{i \in I}$ of nonempty subsets of A such that

P1°. If $\exists x \in A_i \cap A_j$ then $A_i = A_j$ and

P2°. If $x \in A$, then $x \in A_i$ for some $i \in I$.

Examples of partitions are given in the exercises which follow this section.

The results which follow state the connection between equivalence relations in A and partitions of A. They are of great importance in many branches of mathematics.

Let G be an equivalence relation in A; we will sometimes write $x \underset{G}{\sim} y$ instead of $(x, y) \in G$, and say that "*x is equivalent to y modulo G*;" when there is no danger of ambiguity, we will write simply $x \sim y$ and say that "*x is equivalent to y*." Note that since G is an equivalence relation in A, we have

i) $x \sim x, \forall x \in A$.
ii) $x \sim y \Rightarrow y \sim x$.
iii) $x \sim y$ and $y \sim z \Rightarrow x \sim z$.

3.8 Definition Let A be a set and let G be an equivalence relation in A. If $x \in A$, then the *equivalence class of x modulo G* is the set G_x defined as follows:

$$G_x = \{y \in A \mid (y, x) \in G\} = \{y \in A \mid y \underset{G}{\sim} x\}.$$

In other words, G_x is the set of all the elements of A which are equivalent to x. In the mathematical literature, G_x is also denoted by the symbols A_x, $[x]$, x/G.

3.9 Lemma Let G be an equivalence relation in A. Then

$$x \sim y \qquad \text{if and only if} \qquad G_x = G_y.$$

Proof

i) Suppose $x \sim y$; we have

$$z \in G_x \Rightarrow z \sim x \Rightarrow z \sim y \qquad \text{because we assume } x \sim y$$
$$\Rightarrow z \in G_y.$$

We have shown that $G_x \subseteq G_y$; analogously, $G_y \subseteq G_x$; hence $G_x = G_y$.

ii) Suppose $G_x = G_y$; by the reflexive property, $x \sim x$, so $x \in G_x$; but $G_x = G_y$; hence $x \in G_y$, that is, $x \sim y$. ■

3.10 Theorem Let A be a set, let G be an equivalence relation in A, and let $\{G_x\}_{x \in A}$ be the family of all the equivalence classes modulo G. Then

$$\{G_x\}_{x \in A} \text{ is a partition of } A.$$

Proof. By definition, each G_x is a subset of A; it is nonempty because $x \sim x$, hence $x \in G_x$. It remains to prove that P1° and P2° hold.

(P1°) $z \in G_x \cap G_y \Rightarrow z \in G_x$ and $z \in G_y \Rightarrow z \sim x$ and $z \sim y \Rightarrow x \sim z$ and $z \sim y \Rightarrow x \sim y \Rightarrow G_x = G_y$ (the last implication follows by 3.9).

(P2°) If $x \in A$, then by the reflexive property $x \sim x$; hence $x \in G_x$. ■

If G is an equivalence relation in A, and $\{G_x\}_{x \in A}$ is the family of all the equivalence classes modulo G, then $\{G_x\}_{x \in A}$ is referred to as the *partition induced by G*, or the *partition corresponding to G*.

Theorem 3.10 has an important converse, which follows.

3.11 Theorem Let A be a set, let $\{A_i\}_{i \in I}$ be a partition of A, and let G be the set of all pairs (x, y) of elements of A such that x and y are in the same member of the partition; that is,

$$G = \{(x, y) \mid x \in A_i \text{ and } y \in A_i \text{ for some } i \in I\}.$$

Then G is an equivalence relation in A, and $\{A_i\}_{i\in I}$ is the partition induced by G. G is called the *equivalence relation corresponding to* $\{A_i\}_{i\in I}$.

Proof

G is reflexive: $x \in A \Rightarrow x \in A_i$ for some $i \in I \Rightarrow x \in A_i$ and $x \in A_i \Rightarrow (x, x) \in G$.
G is symmetric: $(x, y) \in G \Rightarrow x \in A_i$ and $y \in A_i \Rightarrow y \in A_i$ and $x \in A_i \Rightarrow (y, x) \in G$.
G is transitive: $(x, y) \in G$ and $(y, z) \in G \Rightarrow x \in A_i$ and $y \in A_i$ and $y \in A_j$ and $z \in A_j \Rightarrow A_i = A_j$ (because $y \in A_i \cap A_j$) $\Rightarrow x \in A_i$ and $z \in A_i \Rightarrow (x, z) \in G$.
Finally, each A_i is an equivalence class modulo G; for suppose $x \in A_i$: then $y \in A_i \Leftrightarrow (y, x) \in G \Leftrightarrow y \in G_x$; thus $A_i = G_x$. ■

The last two theorems make it clear that every equivalence relation in A corresponds uniquely to a partition of A, and conversely. Once again: if we are given a partition of A, the *corresponding equivalence relation* is the relation which calls elements x and y "equivalent" if they are in the same member of the partition. Looking at the other side of the coin, if we are given an equivalence relation in A, the *corresponding partition* is the one which puts elements x and y in the same member of the partition iff they are equivalent. The reader should note that G is the equivalence relation corresponding to $\{A_i\}_{i\in I}$ if and only if $\{A_i\}_{i\in I}$ is the partition corresponding to G.

3.12 Example Let $A = \{a, b, c, d, e\}$; let $A_1 = \{a, b\}$, $A_2 = \{c, d\}$ and $A_3 = \{e\}$. Let $G = \{(a, a), (b, b), (c, c), (d, d), (e, e), (a, b), (b, a), (c, d), (d, c)\}$. It is easy to see that $\{A_1, A_2, A_3\}$ is a partition of A (see Fig. 2), and that G is an equivalence relation in A; G is the equivalence relation corresponding to $\{A_1, A_2, A_3\}$, and $\{A_1, A_2, A_3\}$ is the partition corresponding to G. It should be noted that $A_1 = G_a = G_b$, $A_2 = G_c = G_d$, and $A_3 = G_e$.

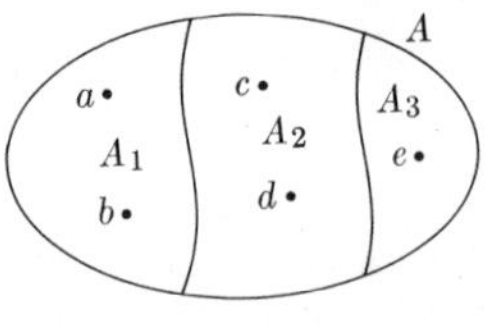

Fig. 2

If G is an equivalence relation in set A, then the set of equivalence classes modulo G is called the *quotient set of A by G*, and is customarily denoted by A/G. Thus, in the preceding example, A/G is the set of three elements $\{G_a, G_c, G_e\}$. The concept of a quotient set plays a vital role in many parts of advanced mathematics.

EXERCISES 3.3

1. Let $\mathbb{Z}$ be the set of the integers. For each integer n, let $B_n = \{m \in \mathbb{Z} \mid \exists q \ni m = n + 5q\}$. Prove that $\{B_n\}_{n \in \mathbb{Z}}$ is a partition of $\mathbb{Z}$.
2. Let $\mathbb{R}$ be the set of the real numbers. In each of the following, prove that $\{B_r\}_{r \in \mathbb{R}}$ is a partition of $\mathbb{R} \times \mathbb{R}$. Describe geometrically the members of this partition. Find the equivalence relation corresponding to each partition.
   a) $B_r = \{(x, y) \mid y = x + r\}$ for each $r \in \mathbb{R}$,
   b) $B_r = \{(x, y) \mid x^2 + y^2 = r\}$ for each $r \in \mathbb{R}$.

   [*Hint*: $y = x + r$ is the equation of a line and $x^2 + y^2 = r$ is the equation of a circle.]
3. Let $\mathbb{R}$ be the set of the real numbers. Prove that each of the following is an equivalence relation in $\mathbb{R} \times \mathbb{R}$:
   a) $G = \{[(a, b), (c, d)] \mid a^2 + b^2 = c^2 + d^2\}$.
   b) $H = \{[(a, b), (c, d)] \mid b - a = d - c\}$.
   c) $J = \{[(a, b), (c, d)] \mid a + b = c + d\}$.

   Find the partition corresponding to each of these equivalence relations, and describe geometrically the members of this partition. [*Hint* for (b): If $b - a = d - c = k$, note that $[(a, b), (c, d)] \in H$ if and only if (a, b) and (c, d) both satisfy the equation $y = x + k$. *Hint* for (c): If $a + b = c + d = k$, note that $[(a, b), (c, d)] \in J$ if and only if (a, b) and (c, d) both satisfy the equation $y = -x + k$.]
4. If H and J are the equivalence relations of Exercise 3, describe the equivalence relation $H \cap J$. Describe the equivalence classes modulo $H \cap J$.
5. Let H and J be the equivalence relations of Exercise 3. Prove that $H \circ J = J \circ H$; conclude that $H \circ J$ is an equivalence relation, and describe the equivalence classes modulo $H \circ J$. [*Hint*: See Exercise 10, Exercise Set 3.2.]
6. Let L be the set of all the straight lines in the plane. Let G and H be the following relations in L:

   $$G = \{(l_1, l_2) \mid l_1 \text{ is parallel to } l_2\}, \qquad H = \{(l_1, l_2) \mid l_1 \text{ is perpendicular to } l_2\}.$$

   Prove the following (argue informally):
   a) G is an equivalence relation in L.
   b) $H \circ G = H$ and $G \circ H = H$.
   c) $G \cup H$ is an equivalence relation; describe its equivalence classes.
7. Let A be an arbitrary set. Prove that I_A and $A \times A$ are equivalence relations in A. Describe the partitions induced, respectively, by I_A and $A \times A$.
8. Let $\{A_i\}_{i \in I}$ be a partition of A and let $\{B_j\}_{j \in J}$ be a partition of B. Prove that $\{A_i \times B_j\}_{(i,j) \in I \times J}$ is a partition of $A \times B$.
9. Suppose $f: A \to B$ is a surjective function, and $\{B_i\}_{i \in I}$ is a partition of B. Prove that $\{\overline{f}(B_i)\}_{i \in I}$ is a partition of A.
10. Suppose $f: A \to B$ is an injective function, and $\{A_i\}_{i \in I}$ is a partition of A. Prove that $\{\overline{f}(A_i)\}_{i \in I}$ is a partition of $\overline{f}(A)$.

11. Let G and H be equivalence relations in A. Prove that each equivalence class modulo $G \cap H$ is the intersection of an equivalence class modulo G with an equivalence class modulo H. More exactly,

$$(G \cap H)_x = G_x \cap H_x, \quad \forall x \ni A.$$

12. Let G and H be equivalence relations in A, and assume that $G \cup H$ is an equivalence relation in A. Prove that each equivalence class modulo $G \cup H$ is the union of an equivalence class modulo G with an equivalence class modulo H. More exactly,

$$(G \cup H)_x = G_x \cup H_x, \quad \forall x \in A.$$

4 PRE-IMAGE, RESTRICTION AND QUOTIENT OF EQUIVALENCE RELATIONS

3.13 Definition Let $f: A \to B$ be a function, and let G be an equivalence relation in B. The *pre-image of G under f* is a relation in A defined as follows:

$$\overline{f}(G) = \{(x, y) \mid (f(x), f(y)) \in G\}.$$

It is simple to show that $\overline{f}(G)$ is an equivalence relation in A.

3.14 Definition Let G be an equivalence relation in A and let $B \subseteq A$. The *restriction of G to B* is a relation in B defined as follows:

$$G_{[B]} = \{(x, y) \mid x \in B \text{ and } y \in B \text{ and } (x, y) \in G\}.$$

It is simple to show that $G_{[B]}$ is an equivalence relation in B.

3.15 Definition Let G and H be equivalence relations in A. We call G a *refinement* of H if $G \subseteq H$; we also say that G is *finer* than H, and that H is *coarser* than G.

3.16 Theorem Let G and H be equivalence relations in A; suppose $G \subseteq H$. Then $z \in H_x \Rightarrow G_z \subseteq H_x$.

Proof. Suppose $z \in H_x$, that is, $(z, x) \in H$; then we have

$$y \in G_z \Rightarrow (y, z) \in G \subseteq H \Rightarrow (y, x) \in H \qquad \text{because we assume } (z, x) \in H$$
$$\Rightarrow y \in H_x.$$

Thus $G_z \subseteq H_x$. ■

3.17 Corollary If $G \subseteq H$, then for each $x \in A$, $G_x \subseteq H_x$.

This follows immediately from 3.16 and the fact that $x \in H_x$.

It follows from 3.16 that **if G is a refinement of H, then each equivalence class modulo H is a union of equivalence classes modulo G.** Indeed, if H_x is an equivalence class modulo H and $z \in H_x$, then, by 3.16, H_x contains the whole class G_z; in other words, H_x contains only *whole* classes modulo G.

3.18 Definition Let G and H be equivalence relations in a set A and let G be a refinement of H. The *quotient of H by G*, which is usually denoted by H/G, is a relation in A/G defined as follows:

$$H/G = \{(G_x, G_y) \mid (x, y) \in H\}.$$

3.19 Theorem H/G is an equivalence relation in A/G.

Proof

H/G is reflexive: For each equivalence class G_x, $(x, x) \in H$ because H is reflexive; thus, by 3.18, $(G_x, G_x) \in H/G$.

H/G is symmetric: $(G_x, G_y) \in H/G \Rightarrow (x, y) \in H \Rightarrow (y, x) \in H \Rightarrow (G_y, G_x) \in H/G$.

H/G is transitive: $(G_x, G_y) \in H/G$ and $(G_y, G_z) \in H/G \Rightarrow (x, y) \in H$ and $(y, z) \in H \Rightarrow (x, z) \in H \Rightarrow (G_x, G_z) \in H/G$. ■

Since H/G is an equivalence relation in A/G, we may write $G_x \underset{H/G}{\sim} G_y$ instead of $(G_x, G_y) \in H/G$. Thus Definition 3.18 may be written in the more suggestive form

3.20 $$G_x \underset{H/G}{\sim} G_y \quad \text{if and only if} \quad x \underset{H}{\sim} y.$$

The reader may easily verify that $G_x \underset{H/G}{\sim} G_y$ if and only if G_x and G_y are subsets of the same equivalence class modulo H.

3.21 Example Let A and G be defined as in Example 3.12; let

$$H = \{(a, a), (b, b), (c, c), (d, d), (e, e), (a, b), (b, a), (c, d), (d, c), (c, e), (e, c), (d, e), (e, d)\}.$$

It is obvious that G is a refinement of H. The partition of A induced by H is $\{H_a, H_c\}$, where $H_a = \{a, b\}$ and $H_c = \{c, d, e\}$ (see Fig. 3). The reader will note that each class modulo H is a union of classes modulo G. Now $A/G = \{G_a, G_c, G_e\}$; by 3.18, H/G is the following relation in A/G:

$$H/G = \{(G_a, G_a), (G_c, G_c), (G_e, G_e), (G_c, G_e), (G_e, G_c)\}.$$

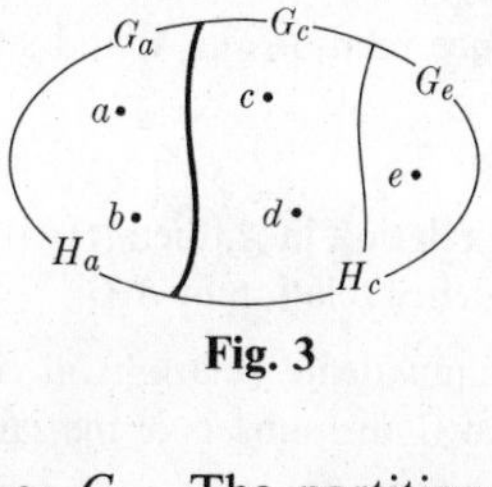

Fig. 3

Note, for instance, that $G_c \underset{H/G}{\sim} G_e$. The partition of A/G induced by H/G is illustrated in Fig. 4. In particular, $(A/G)/(H/G)$ is the set $\{\alpha, \beta\}$, where $\alpha = \{G_a\}$ and $\beta = \{G_c, G_e\}$.

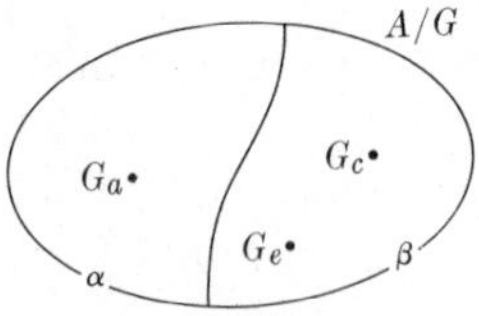

Fig. 4

EXERCISES 3.4

1. Let $A = \{a, b, c, d, e, f\}$, and let G and H be the following equivalence relations in A:
$$G = I_A \cup \{(a, b), (b, a), (b, c), (c, b), (a, c), (c, a), (d, e), (e, d)\},$$
$$H = I_A \cup \{(b, c), (c, b)\}.$$
Clearly H is a refinement of G. Exhibit the sets A/G, A/H, G/H, $(A/H)/(G/H)$.

2. Let $\mathbb{R}$ be the set of the real numbers, and let G be the following relation in $\mathbb{R} \times \mathbb{R}$:
$$G = \{[(a, b), (c, d)] \mid a^2 + b^2 = c^2 + d^2\}.$$
Let $f: \mathbb{R} \to \mathbb{R} \times \mathbb{R}$ be the function given by $f(x) = (\sin x, \cos x)$. Describe $\overline{f}(G)$; what are its equivalence classes?

3. Let $f: A \to B$ be a function and let G be an equivalence relation in B. Prove that $\overline{f}(G)$ is an equivalence relation in A.

4. Let $f: A \to B$ be a function and let G be an equivalence relation in B. Prove that each equivalence class modulo $\overline{f}(G)$ is the inverse image of an equivalence class modulo G. More precisely, if $H = \overline{f}(G)$ and $y = f(x)$, prove that $H_x = \overline{f}(G_y)$.

5. Let G be an equivalence relation in A and suppose that $B \subseteq A$. Prove that $G_{[B]}$ is an equivalence relation in B.

6. Let G be an equivalence relation in A and suppose $B \subseteq A$. Prove that for each $x \in B$, $(G_{[B]})_x = G_x \cap B$.

7. Let G, H, and J be equivalence relations in A, and suppose that $G \subseteq H$ and $H \subseteq J$. Prove that H/G is finer than J/G.

8. Let G, H, and J be equivalence relations in A, and suppose that $G \subseteq H$ and $H \subseteq J$. Prove each of the following.

   a) $G \subseteq H \circ J$.
   b) If $H \circ J$ is an equivalence relation in A, then $(H/G) \circ (J/G) = (H \circ J)/G$.
   c) $(H/G) \circ (J/G)$ is an equivalence relation in A/G.

9. Suppose that G and H are equivalence relations in A, and that $G \subseteq H$. Prove that $G_x \underset{H/G}{\sim} G_y$ if and only if G_x and G_y are subsets of the same equivalence class modulo H.

10. Suppose that G is an equivalence relation in A, and H is an equivalence relation in B.

The *product* of G and H is defined to be the following relation in $A \times B$:

$$G \bullet H = \{[(x, w), (y, z)] \mid (x, y) \in G \text{ and } (w, z) \in H\}.$$

Prove that $G \bullet H$ is an equivalence relation in $A \times B$.

11. Prove that every equivalence relation in a set A is the pre-image of an equivalence relation in $A \times A$. [*Hint:* Let $f: A \to A \times A$ be the function given by $f(x) = (x, x)$; if G is a relation in A, consider the relation $G \bullet G$ (see Exercise 10) in $A \times A$.]

12. Let $f: A \to B$ be a function and let G be an equivalence relation in B. Prove that $\overline{f}(G) = f^{-1} \circ G \circ f$.

5 EQUIVALENCE RELATIONS AND FUNCTIONS

If $f: A \to B$ is a function, we define a relation G in A as follows:

$$G = \{(x, y) \mid f(x) = f(y)\}.$$

It is easy to see that G is an equivalence relation in A. G is called the *equivalence relation determined by f*.

Conversely, if G is an equivalence relation in a set A, we define a function $f: A \to A/G$ as follows:

$$f(x) = G_x, \quad \forall x \in A.$$

It is easy to see that f is a function; f is called the *canonical function from A to A/G*.

3.22 Theorem Let G be an equivalence relation in a set A. If f is the canonical function from A to A/G, then G is the equivalence relation determined by f.

Proof. Let f be the canonical function from A to A/G, and let H be the equivalence relation determined by f; we will prove that $G = H$:

$$(x, y) \in G \Leftrightarrow G_x = G_y \Leftrightarrow f(x) = f(y) \Leftrightarrow (x, y) \in H. \quad \blacksquare$$

Let A and B be sets and let $f: A \to B$ be a function; we will define three functions r, s, t, obtained from f, which play an important role in many mathematical arguments. Let G be the equivalence relation determined by f:

$r: A \to A/G$ is the canonical function from A to A/G.

$s: A/G \to \overline{f}(A)$ is the function given by $s(G_x) = f(x)$, $\forall x \in A$.

$t: \overline{f}(A) \to B$ is the function given by $t(y) = y$, $\forall y \in \overline{f}(A)$.

Note that t is the inclusion function of $\overline{f}(A)$ in B (see 2.12).

3.23 Theorem Let A and B be sets, let $f: A \to B$ be a function, let G be the equivalence relation determined by f, and let r, s, t be the functions defined above. Then r is surjective, s is bijective, t is injective, and $f = t \circ s \circ r$.

Proof

i) If $G_x \in A/G$, then $x \in A$ and $r(x) = G_x$; thus r is surjective.

ii) If $f(x) \in \overline{f}(A)$, then $x \in A$, $G_x \in A/G$, and $f(x) = s(G_x)$; thus s is surjective.

iii) $s(G_x) = s(G_y) \Rightarrow f(x) = f(y) \Rightarrow (x, y) \in G \Rightarrow G_x = G_y$; thus s is injective.

iv) $t(y_1) = t(y_2) \Rightarrow y_1 = y_2$; thus t is injective.

v) Let $x \in A$; $t\{s[r(x)]\} = t[s(G_x)] = t(f(x)) = f(x)$; thus

$$[t \circ s \circ r](x) = f(x), \forall x \in A,$$

so by 2.5, $t \circ s \circ r = f$. ■

We may sum up the foregoing results by saying that any function $f: A \to B$ can be expressed as a composite of three functions r, s, t which are, respectively, surjective, bijective, and injective. This is referred to as the *canonical decomposition* of f, and it is customarily exhibited in a diagram such as the following:

$$A \xrightarrow[\text{surj}]{r} A/G \xrightarrow[\text{bij}]{s} \overline{f}(A) \xrightarrow[\text{inj}]{t} B.$$

One of the results of 3.23 is especially useful; namely, that if $f: A \to B$ is a function and G is the equivalence relation determined by f, then A/G and $\overline{f}(A)$ are in one-to-one correspondence. This is customarily expressed by writing $A/G \approx \overline{f}(A)$. In particular,

3.24 if f is surjective, then $A/G \approx B$.

Let A and B be sets, let $f: A \to B$ be a function, and let H be the equivalence relation determined by f. Let G be any equivalence relation in A which is finer than H. We define a function from A/G to B as follows:

3.25 $$[f/G](G_x) = f(x), \quad \forall x \in A.$$

It is easy to see that f/G is a function from A/G to B; f/G is called the *quotient* of f by G.

3.26 Theorem Let $f: A \to B$ be a function, let H be the equivalence relation determined by f, and let G be a refinement of H. Then H/G is the equivalence relation determined by f/G.

Proof. Let J be the equivalence relation determined by f/G; we will prove that $J = H/G$. Indeed,

$$\begin{aligned}(G_x, G_y) \in J &\Leftrightarrow [f/G](G_x) = [f/G](G_y)\\ &\Leftrightarrow f(x) = f(y)\\ &\Leftrightarrow (x, y) \in H\\ &\Leftrightarrow (G_x, G_y) \in H/G. \blacksquare\end{aligned}$$

As an example of the use of Theorem 3.26, consider the following situation: G and H are equivalence relations in A, $G \subseteq H$, f is the canonical function from A to A/H (hence, by 3.22, H is the equivalence relation determined by f). Thus, by 3.25, f/G is a function from A/G to A/H, and by 3.26, H/G is the equivalence relation determined by f/G. It is easy to see that f/G is surjective, because f is surjective. Therefore, by 3.24,

$$(A/G)/(H/G) \approx A/H.$$

EXERCISES 3.5

1. Let $f: A \to B$ be a surjective function, let G be the equivalence relation induced by f, and let H be an equivalence relation in A which is coarser than G. Define the *image of H* as follows:

$$\overline{f}(H) = \{(f(x), f(y)) \backslash (x, y) \in H\}$$

Prove that $\overline{f}(H)$ is an equivalence relation in B.

2. Let $f: A \to B$ be a surjective function, and let G be the equivalence relation induced by f. Let J be any equivalence relation in B. Prove that

   a) $\overline{f}(J)$ is coarser than G,

   b) $H = \overline{f}(J)$ if and only if $J = \overline{f}(H)$. (See Exercise 1 above.)

   Conclude that there exists a one-to-one correspondence between the equivalence relations in B and the equivalence relations in A which are coarser than G.

3. Let $f: A \to B$ be a function, and let G be the equivalence relation determined by f. Prove that $G = f^{-1} \circ f$.

4. Let $f: A \to B$ and $g: B \to C$ be functions, and let G be the equivalence relation determined by g. Prove that $\overline{f}(G)$ is the equivalence relation determined by $g \circ f$.

5. Let G and H be equivalence relations in a set A, and suppose that $G \subseteq H$. Let f be the canonical function from A to A/G. Prove that $H/G = \overline{f}(H)$. (See Exercise 1.)

6. Let G and H be equivalence relations in A, and suppose that $G \subseteq H$. Let f be the canonical function from A to A/G, and let g be the canonical function from A to A/H. Let $h = g/G$. Prove that $g = h \circ f$.

7. Let G and H be equivalence relations in A, and suppose that $G \subseteq H$. If f is the canonical function from A to A/G, prove that $H = \overline{f}(H/G)$.

8. Let G, H, and J be equivalence relations in A and suppose that $G \subseteq H \subseteq J$. Let $f: A \to A/G$, $g: A \to A/H$, and $h: A \to A/J$ be the canonical functions associated, respectively, with G, H, and J. Prove that $h/G = h/H \circ g/G$.

9. Let G and H be equivalence relations in A, and suppose that $G \subseteq H$. Let f be the canonical function from A to A/H. Prove that f/G is surjective.

10. Let G and H be arbitrary equivalence relations in A. Prove that

    a) $A/(G \circ H) \approx (A/G)/(G \circ H/G)$. b) $A/G \approx (A/G \cap H)/(G/G \cap H)$.

11. Let $f: A \to A$ be a function, and let G be the equivalence relation determined by f. Prove that $f \circ f = f$ if and only if

$$z \in G_x \Rightarrow f(z) \in G_x,$$

for every $z, x \in A$.

4
Partially Ordered Classes

1 FUNDAMENTAL CONCEPTS AND DEFINITIONS

By a *partially ordered class* we mean a pair* of objects $\langle A, G\rangle$, where A is a class and G is an order relation in A. We say that A is ordered by G, or that G orders A. If A is a set, we say that $\langle A, G\rangle$ is a *partially ordered set*.

In ordinary mathematical applications, every partially ordered class is a partially ordered set. However, the intuitive idea of an "ordered collection of elements" is meaningful for any collection A, whether A be a set or a proper class; hence it is natural to give the definition in its most general form, letting A be any class. Once again, since every set is a class, everything we have to say about partially ordered classes applies, in particular, to partially ordered sets.

Let $\langle A, G\rangle$ be a partially ordered class; if it is well understood, in a given discussion, that G is the order relation in A, then we will say loosely that *A is a partially ordered class*. If A is a partially ordered class, ordered by G, it is customary to write $x \leqslant y$ to denote the fact that $(x, y) \in G$. We further agree that $y \geqslant x$ has the same meaning as $x \leqslant y$, and that $x \nleqslant y$ means that $(x, y) \notin G$.

If $x \in A$ and $y \in A$ and $x \leqslant y$, then we say that "x is less than or equal to y." We agree that $x < y$ is an abbreviation for "$x \leqslant y$ and $x \neq y$." If $x < y$, we say that "x is strictly less than y." Note that $<$ is *not* an order relation; it is irreflexive, asymmetric, and transitive.

If A is a partially ordered class and B is a subclass of A, we may consider B to be ordered by the order relation in A. Specifically, if $x \in B$ and $y \in B$, then we let $x \leqslant y$ in B if and only if $x \leqslant y$ in A.

If A and B are partially ordered classes, there are several possible ways of ordering the class $A \times B$; the two most useful ways of doing so are given in the following definitions.

4.1 Definition Let A and B be partially ordered classes; by the *lexicographic ordering* of $A \times B$ we mean the following order relation in $A \times B$: If

---

* If A and G are not both sets, the ordered pair $\langle A, G\rangle$ may be defined formally thus:

$$\langle A, G\rangle = (A \times \{\varnothing\}) \cup (G \times \{\{\varnothing\}\}).$$

$(a_1, b_1) \in A \times B$ and $(a_2, b_2) \in A \times B$, then we let $(a_1, b_1) \leqslant (a_2, b_2)$ if and only if

i) $a_1 < a_2$ or
ii) $a_1 = a_2$ and $b_1 \leqslant b_2$.

The lexicographic ordering is so called because it imitates the way we order words in the dictionary (for example, *be* precedes *go* because *b* precedes *g*, and *be* precedes *by* because *e* precedes *y*).

4.2 Definition Let A and B be partially ordered classes; by the *antilexicographic ordering* of $A \times B$ we mean the following order relation in $A \times B$: If $(a_1, b_1) \in A \times B$ and $(a_2, b_2) \in A \times B$, then we let $(a_1, b_1) \leqslant (a_2, b_2)$ if and only if

i) $b_1 < b_2$ or
ii) $b_1 = b_2$ and $a_1 \leqslant a_2$.

4.3 Definition Let A be a partially ordered class. Two elements x and y in A are said to be *comparable* if either $x \leqslant y$ or $y \leqslant x$; otherwise, they are said to be *incomparable*.

4.4 Definition Let A be a partially ordered class, and let B be an arbitrary subclass of A. If every two elements of B are comparable, then we call B a *fully ordered subclass* of A, or a *linearly ordered* subclass of A, or, more commonly, a *chain* of A. If every two elements of A are comparable, then A is called a *fully ordered*, or *linearly ordered*, class.

4.5 Definition Let A be a partially ordered class and suppose $a \in A$. The *initial segment of A determined by a* is the class S_a, defined as follows:

$$S_a = \{x \in A \mid x < a\}.$$

4.6 Theorem Let A be a partially ordered class. If P is an initial segment of A, and Q is an initial segment of P, then Q is an initial segment of A.

Proof. By hypothesis, $P = \{x \in A \mid x < a\}$ for some $a \in A$, and $Q = \{x \in P \mid x < b\}$ for some $b \in P$. Let $Q_1 = \{x \in A \mid x < b\}$; Q_1 is obviously an initial segment of A; we will show that $Q = Q_1$, and the theorem will thus be proved. Clearly, $Q \subseteq Q_1$; conversely, if $x \in Q_1$, then $x \in A$ and $x < b$; but $b < a$ because $b \in P$; hence $x < a$, and it follows that $x \in P$; thus $x \in Q$. ■

Theorem 4.6 may be paraphrased as follows:

An initial segment of an initial segment of *A* is an initial segment of *A*.

4.7 Definition If A is a partially ordered class, then a *cut* of A is a pair (L, U) of nonempty subclasses of A with the following properties:

i) $L \cap U = \varnothing$ and $L \cup U = A$.

ii) If $x \in L$ and $y \leqslant x$, then $y \in L$.

iii) If $x \in U$ and $y \geqslant x$, then $y \in U$.

It is convenient to use a graphic device called a *line diagram* to illustrate simple properties of partially ordered classes. The elements of the class are represented by points on the diagram; if two points x and y are connected by a line, and the line rises from x to y, this means that $x \leqslant y$.

4.8 Example Fig. 1 represents a partially ordered class with six elements. Note that $\{a, b, c\}$ and $\{d, e, b, c\}$ are chains of A; $S_e = \{b, c, f\}$ is the initial segment determined by e; if $L = \{a, b, c\}$ and $U = \{d, e, f\}$, then (L, U) is a cut.

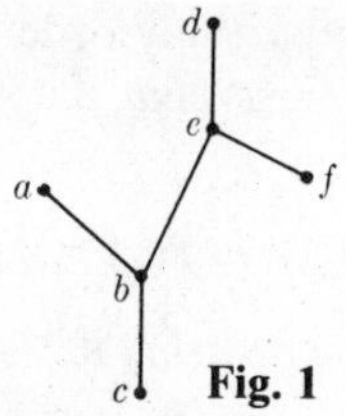

Fig. 1

4.9 Example The most important order relation in mathematics is the class inclusion relation $\subseteq$; it is reflexive, for if A is any class, then $A \subseteq A$; it is antisymmetric, because if $A \subseteq B$ and $B \subseteq A$, then $A = B$; it is transitive, for $A \subseteq B$ and $B \subseteq C$ imply that $A \subseteq C$. If $\mathscr{A}$ is a class (of classes) and we consider $\mathscr{A}$ to be ordered by the inclusion relation, we say that $\mathscr{A}$ is *ordered by inclusion.* Note that if $\mathscr{C}$ is a chain of $\mathscr{A}$, this means that for any two elements $A, B \in \mathscr{C}$, either $A \subseteq B$ or $B \subseteq A$.

EXERCISES 4.1

1. Let $A = \{a, b, c, d\}$; if $\mathscr{P}(A)$ is ordered by inclusion, draw its line diagram.
2. Let A be the partially ordered class defined by the following diagram.

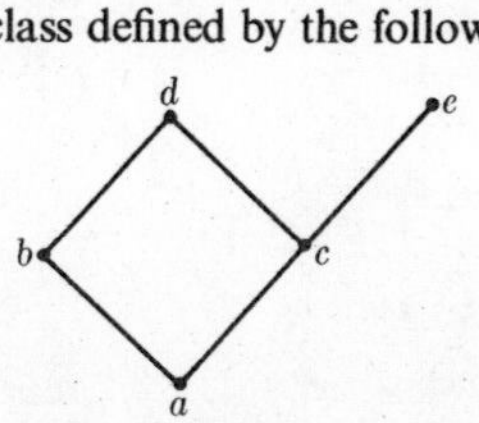

List all the chains of A, all the initial segments of A, and all the cuts of A.

3. Let A be the partially ordered class of Exercise 2. Draw the line diagram for the following classes: the class of all the chains of A (ordered by inclusion), the class of all the initial segments of A (ordered by inclusion), the class of all the cuts of A (ordered by inclusion on the "left component" L).
4. Let A and B be partially ordered classes, let C be a chain of A, and let D be a chain of B. If $A \times B$ is ordered lexicographically (4.1), prove that $C \times D$ is a chain of $A \times B$.
5. Let A and B be partially ordered classes, and let $A \times B$ be ordered antilexicographically (4.2). Prove that if (L, U) is a cut of B, then $(A \times L, A \times U)$ is a cut of $A \times B$.
6. Let A be a partially ordered class, and let G be an equivalence relation in A. Suppose the following condition holds: If $x \underset{G}{\sim} z$ and $x \leqslant y \leqslant z$, then $y \underset{G}{\sim} z$. Define a relation H in A/G by $$H = \{(G_x, G_y) \mid \exists w \in G_x, \exists z \in G_y \ni w \leqslant z\}.$$ Prove that H is an order relation in A/G.

2 ORDER PRESERVING FUNCTIONS AND ISOMORPHISM

4.10 Definition Let A and B be partially ordered classes; a function $f: A \to B$ is said to be *increasing*, or *order-preserving*, if it satisfies the following condition: For every two elements $x, y \in A$,

$$x \leqslant y \Rightarrow f(x) \leqslant f(y).$$

We say that $f: A \to B$ is *strictly increasing* if it satisfies the following condition: For every two elements $x \in A$ and $y \in A$,

$$x < y \Rightarrow f(x) < f(y).$$

4.11 Definition Let A and B be partially ordered classes; a function $f: A \to B$ is called an *isomorphism* if it is bijective and satisfies the following condition: For every two elements $x \in A$ and $y \in A$,

$$x \leqslant y \Leftrightarrow f(x) \leqslant f(y).$$

Figures 2, 3, and 4 provide simple illustrations of the concepts we have

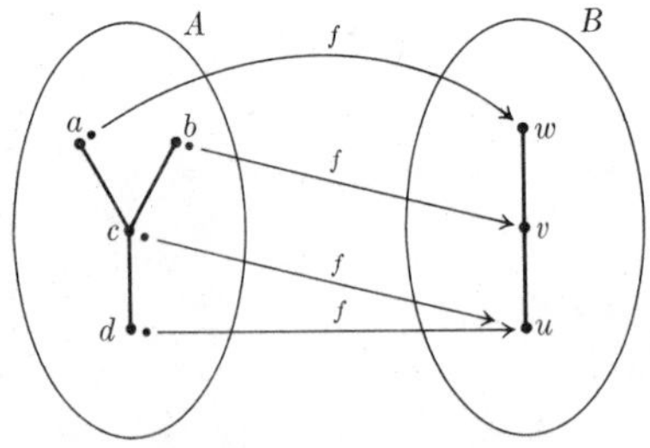

$f: A \to B$ is an increasing function

Fig. 2

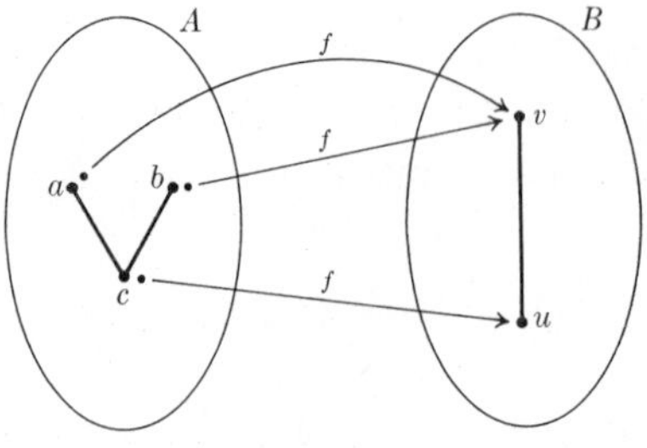

$f: A \to B$ is a strictly increasing function

Fig. 3

just defined. Figure 5 describes a function which is bijective and increasing, but *is not an isomorphism* [note that $f(b) < f(a)$ but a and b are incomparable]. The reader should compare this example with Definition 4.11.

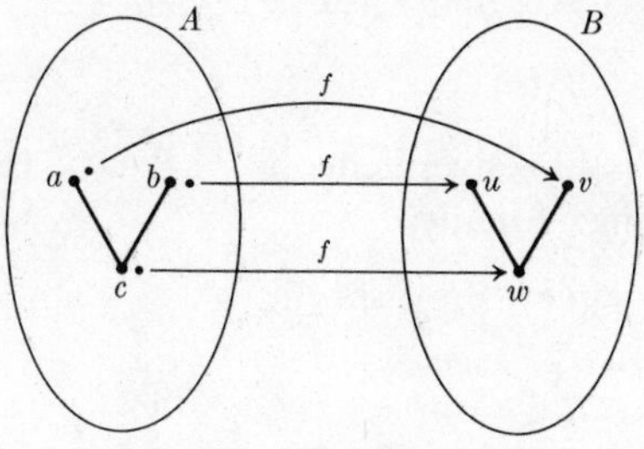

$f\colon A \to B$ is an isomorphism

Fig. 4

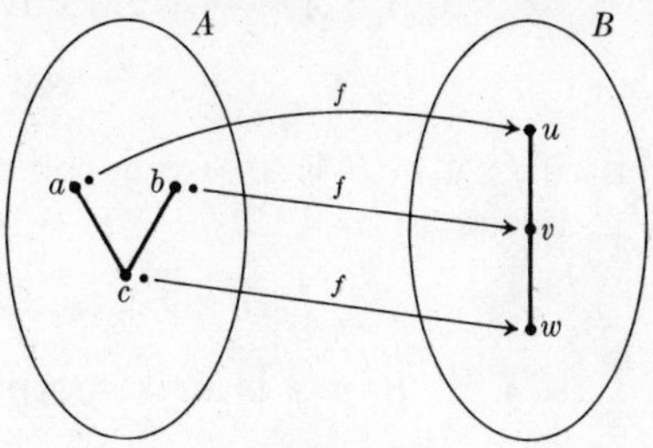

Fig. 5

4.12 Theorem If $f\colon A \to B$ is an isomorphism, then

$$x < y \Leftrightarrow f(x) < f(y).$$

Proof

i) Let us assume that $x < y$; then $x \leqslant y$; hence $f(x) \leqslant f(y)$. If $f(x) = f(y)$, then $x = y$, which is contrary to our assumption; thus $f(x) < f(y)$.

ii) The converse is proved by the same argument. ■

4.13 Theorem Let A and B be partially ordered classes and let $f\colon A \to B$ be a bijective function. Then $f\colon A \to B$ is an isomorphism if and only if $f\colon A \to B$ and $f^{-1}\colon B \to A$ are increasing functions.

Proof. Note, first, that if f is bijective, then $\forall x \in A$, $f^{-1}(f(x)) = x$. Now suppose that f and f^{-1} are increasing functions:

$$f(x) \leqslant f(y) \Rightarrow f^{-1}(f(x)) \leqslant f^{-1}(f(y)) \Rightarrow x \leqslant y;$$

from this, and the fact that f is increasing, we deduce that f is an isomorphism. Conversely, if f is an isomorphism, then certainly f is increasing; furthermore, if $f(x)$ and $f(y)$ are arbitrary elements of B, then

$$f(x) \leqslant f(y) \Rightarrow x \leqslant y \Rightarrow f^{-1}(f(x)) \leqslant f^{-1}(f(y));$$

so f^{-1} is increasing. ■

4.14 Theorem Let A, B, and C be partially ordered classes.

i) The identity function $I_A\colon A \to A$ is an isomorphism.

ii) If $f\colon A \to B$ is an isomorphism, then $f^{-1}\colon B \to A$ is an isomorphism.

iii) If $f: A \to B$ and $g: B \to C$ are isomorphisms, then $g \circ f: A \to C$ is an isomorphism.

Proof

i) By 2.10, $I_A: A \to A$ is bijective; now $I_A(x) = x$ and $I_A(y) = y$; hence

$$x \leqslant y \Leftrightarrow I_A(x) \leqslant I_A(y).$$

ii) If $f: A \to B$ is an isomorphism, then it is bijective, hence $f^{-1}: B \to A$ is bijective; by 4.13, $f^{-1}: B \to A$ is increasing. Finally,

$$f^{-1}(x) \leqslant f^{-1}(y) \Rightarrow f(f^{-1}(x)) \leqslant f(f^{-1}(y)) \Rightarrow x \leqslant y,$$

so $f^{-1}: B \to A$ is an isomorphism.

The proof of (iii) is left as an exercise for the reader. ■

4.15 Definition If A and B are partially ordered classes and there exists an isomorphism from A to B, we say that *A is isomorphic with B.*

Theorem 4.14 indicates that the relation "A is isomorphic with B" is an equivalence relation among partially ordered classes. Indeed, by 4.14(i), A is isomorphic with A; by 4.14(ii), if A is isomorphic with B, then B is isomorphic with A; by 4.14(iii), if A is isomorphic with B and B is isomorphic with C, then A is isomorphic with C.

We will write $A \approxeq B$ to denote the fact that A is isomorphic with B.

The concept of isomorphism is of great importance in the study of partially ordered classes. Suppose that A and B are partially ordered classes and $f: A \to B$ is an isomorphism; let us agree to write x' instead of $f(x)$; since $f: A \to B$ is bijective, every element x in A corresponds with a unique element x' in B:

$$x \overset{f}{\mapsto} x',$$

$$y \overset{f}{\mapsto} y',$$

$$z \overset{f}{\mapsto} z', \quad \text{etc.....}$$

Furthermore, by 4.12, if x and y are any two elements in A, then

$$x < y \Leftrightarrow x' < y'.$$

This means that the ordering of A is exactly the same as the ordering of B; in particular (if it is practical to draw the line diagrams of A and B), the line diagram of A is the same as the line diagram of B. Thus, essentially, there is no difference between A and B except the letters we use to designate their elements.

4.16 Example If $B = \{l, m, n, o, p, q\}$ is ordered as in the following diagram,

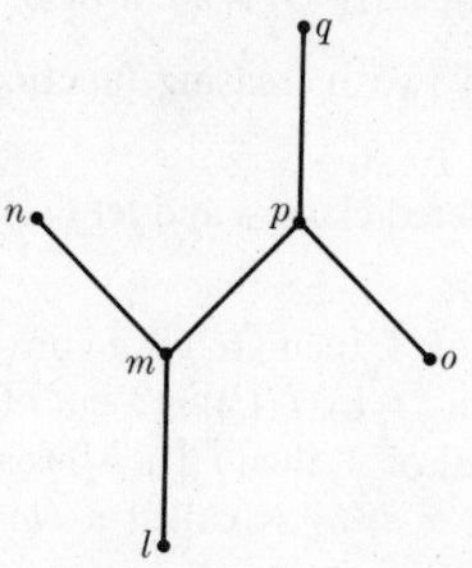

then B is isomorphic with the class A of Example 4.8. The isomorphism $f: A \to B$ is given by the following table.

4.17

x	$f(x)$
a	n
b	m
c	l
d	q
e	p
f	o

In conclusion, if A is isomorphic with B, and if we identify corresponding elements, then A and B are essentially the same partially ordered class.

EXERCISES 4.2

1. Prove that if $f: A \to B$ is an injective, increasing function, then it is strictly increasing.
2. Let $f: A \to B$ be an increasing function. If C is a chain of A, prove that $\overline{f}(C)$ is a chain of B.
3. Let A and B be partially ordered classes. Prove that if $A \times B$ is ordered lexicographically, then the projection function $p_1: A \times B \to A$ is increasing; if $A \times B$ is ordered antilexicographically, then the projection function $p_2: A \times B \to B$ is increasing. [Note that $p_1(x, y) = x$, $p_2(x, y) = y$.]
4. A subclass C of a partially ordered class is called *convex* if it satisfies the following condition: If $a \in C$ and $b \in C$ and $a \leqslant x \leqslant b$, then $x \in C$. Let A and B be partially ordered classes, let $f: A \to B$ be an increasing function, and let C be a convex subclass of B. Prove that $\overline{f}(C)$ is a convex subclass of A.

5. Let A and B be partially ordered classes, let $f: A \to B$ be an increasing function, and let H be the equivalence relation determined by f. Prove that each equivalence class modulo H is a convex subclass of A.

6. Let A and B be partially ordered classes, and let $f: A \to B$ be an increasing function; assume $\overline{f}(A) = B$. Prove that if (L, U) is a cut of B, then $(\overline{f}(L), \overline{f}(U))$ is a cut of A.

7. Prove that the composite of two increasing functions is increasing. Use this result to prove 4.14(iii).

8. Let A and B be partially ordered classes, and let $f: A \to B$ be an isomorphism. Prove each of the following:

   a) If C is a convex subclass of A, then $\overline{f}(C)$ is a convex subclass of B.
   b) If (L, U) is a cut of A, then $(\overline{f}(L), \overline{f}(U))$ is a cut of B.
   c) If $[a, b]$ is a closed interval of A, then $\overline{f}([a, b])$ is a closed interval of B. (If $a, b \in A$, then the set $\{x \in A \mid a \leqslant x \leqslant b\}$ is called a *closed interval* of A and is denoted by the symbol $[a, b]$.)

9. Let E and F be partially ordered classes, and let $g: E \to F$ be an isomorphism. Prove that for arbitrary $x \in E$, $\bar{g}(S_x) = S_{g(x)}$; conclude that $S_x \approxeq S_{g(x)}$.

10. Let A be a partially ordered set. For each $a \in A$, let $I_a = \{x \in A \mid x \leqslant a\}$. Let $\mathscr{I} = \{I_a\}_{a \in A}$ and let $\mathscr{I}$ be ordered by inclusion. Prove that $\mathscr{I}$ is isomorphic with A.

11. Let A, B, and C be mutually disjoint, partially ordered classes. If $A \approxeq B$, prove that $(A \cup C) \approxeq (B \cup C)$. [*Hint:* Take the union of two functions, as in 2.16.]

12. Let A be a partially ordered set. Define $L_x = \{z \in A \mid z \leqslant x\}$ and $U_x = \{z \in A \mid z \nleqslant x\}$. Prove the following:

    a) For each $x \in A$, (L_x, U_x) is a cut of A.
    b) The function ϕ defined by $\phi(x) = (L_x, U_x)$ is an isomorphism between A and the class of all the cuts of the above-described form.

3 DISTINGUISHED ELEMENTS. DUALITY

Certain distinguished elements play an important part in the study of partially ordered classes. We now define them; in each of the following definitions, we assume that A is a partially ordered class.

4.18 Definition An element $m \in A$ is called a *maximal element* of A if none of the elements of A are strictly greater than m; in symbols, this can be expressed as follows:

$$\forall x \in A, \quad \text{if } x \geqslant m, \qquad \text{then} \quad x = m.$$

Similarly, an element $n \in A$ is called a *minimal element* of A if none of the elements of A are strictly less than n; in symbols,

$$\forall x \in A, \quad \text{if } x \leqslant n, \qquad \text{then} \quad x = n.$$

4.19 Definition An element $a \in A$ is called the *greatest element* of A if $a \geqslant x$ for every $x \in A$. An element $b \in A$ is called the *least element* of A if $b \leqslant x$ for every $x \in A$.

It is easy to see that if A has a greatest element, then this element is unique; for suppose a and a' are both greatest elements of A. Then $a \leqslant a'$ and $a' \leqslant a$; hence $a = a'$. Analogously, the least element of A is unique.

4.20 Definition Let B be a subclass of A. An *upper bound of B in A* is an element $a \in A$ such that $a \geqslant x$ for every $x \in B$. A *lower bound of B in A* is an element $b \in A$ such that $b \leqslant x$ for every $x \in B$. When there is no risk of ambiguity, we will refer to an upper bound of B in A simply as an "*upper bound of B*," and to a lower bound of B in A simply as a "*lower bound of B*." The class of all the upper bounds of B will be denoted by $\upsilon(B)$ and the class of all the lower bounds of B will be denoted by $\lambda(B)$.

4.21 Definition If the class of lower bounds of B in A has a greatest element, then this element is called the *greatest lower bound* of B in A. If the class of upper bounds of B in A has a least element, then this element is called the *least upper bound* of B in A. The least upper bound of B in A is also called the *supremum* of B in A (abbreviated $\sup_A B$), and the greatest lower bound of B in A is also called the *infimum* of B in A (abbreviated $\inf_A B$). When there is no risk of ambiguity, we will write $\sup B$ for $\sup_A B$, and $\inf B$ for $\inf_A B$.

We have seen that the greatest element and the least element of any class are unique; hence the sup and the inf, if they exist, are unique.

Examples

4.22 Figure 6 is the line diagram of a class that has maximal elements but no greatest element (a and d are maximal elements).

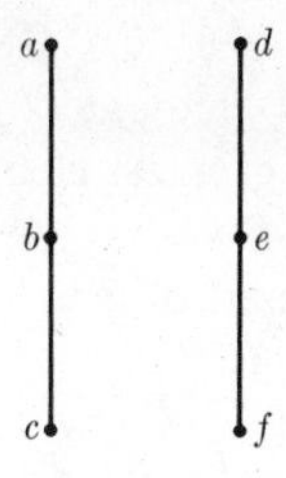

Fig. 6

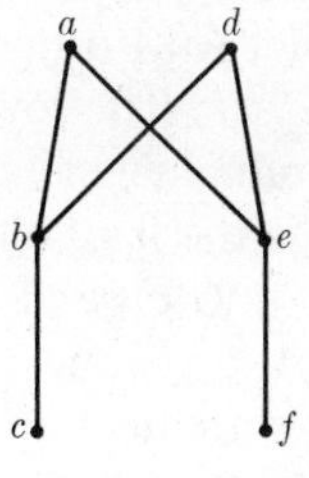

Fig. 7

4.23 In Fig. 7, let $A = \{a, b, c, d, e, f\}$ and let $B = \{b, c, e, f\}$. B has two upper bounds in A, namely a and d, but no sup.

4.24 In Fig. 7, let A and B be defined as above, and let $C = \{a, b, c, e, f\}$ and $D = \{d, b, c, e, f\}$. Then B has no sup in A, although $\sup_C B = a$ and $\sup_D B = d$.

4.25 The class N of all the positive integers has a least element but no greatest element and no maximal elements. The class $\mathbb{Z}$ of all the integers has neither a greatest nor a least element.

4.26 Let $\mathscr{A}$ be a class (of classes) which is ordered by inclusion; let $\mathscr{B} = \{B_i\}_{i \in I}$ be a subclass of $\mathscr{A}$, and let us assume that $\bigcup_{i \in I} B_i$ and $\bigcap_{i \in I} B_i$ are elements of $\mathscr{A}$. Then $\sup \mathscr{B} = \bigcup_{i \in I} B_i$: indeed, each B_i is $\subseteq \bigcup_{i \in I} B_i$, hence $\bigcup_{i \in I} B_i$ is an upper bound of $\mathscr{B}$; furthermore, if C is any other upper bound of $\mathscr{B}$, this means that $B_i \subseteq C$ for every $i \in I$; hence, by 1.40(i), $\bigcup_{i \in I} B_i \subseteq C$; this proves that $\bigcup_{i \in I} B_i$ is the least upper bound of $\mathscr{B}$. Similarly, $\inf \mathscr{B} = \bigcap_{i \in I} B_i$, for clearly $\bigcap_{i \in I} B_i$ is $\subseteq$ each B_i, hence $\bigcap_{i \in I} B_i$ is a lower bound of $\mathscr{B}$; furthermore, if D is any other lower bound of $\mathscr{B}$, this means that $D \subseteq B_i$ for every $i \in I$, hence, by 1.40(ii), $D \subseteq \bigcap_{i \in I} B_i$; this proves that $\bigcap_{i \in I} B_i$ is the greatest lower bound of $\mathscr{B}$.

4.27 It is important to note that $\lambda(\varnothing) = A$. Indeed, if $x \in A$, then the statement "$x \leqslant y$ for every $y \in \varnothing$" is not false (for to deny it would be to assert that "$\exists y \in \varnothing \ni x \nleqslant y$," which is absurd); hence it is true. This holds for each element $x \in A$, so we conclude that $\lambda(\varnothing) = A$. Next, we note that $\inf \varnothing$ is the greatest element of $\lambda(\varnothing)$; thus if A has a greatest element a, then $a = \inf \varnothing$. Analogously, if b is the least element of A, then $b = \sup \varnothing$.

When we reason about partially ordered classes, we are led to the interesting notion of *duality*. Briefly, duality can be explained as follows.

If G is an order relation in A, then G^{-1} is also an order relation in A (see Exercise 8, Exercise Set 3.2). Let $\langle A, G\rangle$ refer to the class A ordered by G, and let $\langle A, G^{-1}\rangle$ refer to A ordered by G^{-1}. Then $x \leqslant y$ in $\langle A, G\rangle$ if and only if $x \geqslant y$ in $\langle A, G^{-1}\rangle$; it follows that a is a maximal element of $\langle A, G\rangle$ if and only if a is a minimal element of $\langle A, G^{-1}\rangle$; a is the greatest element of $\langle A, G\rangle$ if and only if a is the least element of $\langle A, G^{-1}\rangle$; if $B \subseteq A$, then b is an upper bound of B in $\langle A, G\rangle$ if and only if b is a lower bound of B in $\langle A, G^{-1}\rangle$; and $b = \sup B$ in $\langle A, G\rangle$ if and only if $b = \inf B$ in $\langle A, G^{-1}\rangle$.

Let $\mathscr{S}$ be a statement about partially ordered classes. In $\mathscr{S}$, suppose that we replace each occurrence of $\leqslant$ by $\geqslant$ and vice versa; suppose furthermore that we replace the words "maximal" by "minimal" and vice versa, "greatest" by "least" and vice versa, "upper bound" by "lower bound" and vice versa, "sup" by "inf" and vice versa. The resulting statement $\mathscr{S}'$ is called the *dual* of $\mathscr{S}$.

In view of what we have said above, it is easy to see that if $\mathscr{S}$ is a true statement in $\langle A, G^{-1}\rangle$, then the dual of $\mathscr{S}$ is true in $\langle A, G\rangle$. In particular, suppose that $\mathscr{S}$ is a theorem for all partially ordered classes. If A is any partially ordered class and G is the order relation in A, then $\mathscr{S}$ is true in $\langle A, G^{-1}\rangle$, hence the dual of $\mathscr{S}$ is true in $\langle A, G\rangle$. Thus if $\mathscr{S}$ is a theorem for all partially ordered classes, the dual of $\mathscr{S}$ is also a theorem.

The concept of duality permits a considerable economy in the presentation of theorems about partially ordered classes, for every time we prove a theorem, we know that the dual of the theorem is also true.

The following are a few properties of the distinguished elements in a partially ordered class.

4.28 Theorem If A has a greatest element a, and B has a greatest element b, and $A \subseteq B$, then $a \leqslant b$.

Proof. By definition, $b \geqslant x$ for every $x \in B$; but $a \in A \subseteq B$; hence $b \geqslant a$. ■

Dual If A has a least element a, and B has a least element b, and $A \subseteq B$, then $a \geqslant b$.

4.29 Theorem Let B and C be subclasses of A. If $B \subseteq C$, then $\upsilon(C) \subseteq \upsilon(B)$.

Proof. $x \in \upsilon(C) \Rightarrow x \geqslant y, \forall y \in C \Rightarrow x \geqslant y, \forall y \in B \Rightarrow x \in \upsilon(B)$. ■

Dual If $B \subseteq C$, then $\lambda(C) \subseteq \lambda(B)$.

4.30 Theorem Let B and C be subclasses of A, and suppose that B and C each has a sup in A. If $B \subseteq C$ then $\sup B \leqslant \sup C$.

Proof. By 4.29, $\upsilon(C) \subseteq \upsilon(B)$; hence by 4.28 (dual), $\sup B \leqslant \sup C$. ■

Dual If B and C each has a inf in A, and if $B \subseteq C$, then $\inf B \geqslant \inf C$.

4.31 Theorem Let B be a subclass of A. Then $B \subseteq \upsilon(\lambda(B))$.

Proof. Suppose $x \in B$; for each $y \in \lambda(B)$, $y \leqslant x$, that is, $x \geqslant y$; hence $x \in \upsilon(\lambda(B))$. ■

Dual $B \subseteq \lambda(\upsilon(B))$.

4.32 Lemma Let B be a subclass of A and suppose that $\lambda(B)$ has a sup in A. Then B has an inf in A, and $\inf B = \sup \lambda(B)$.

Proof. Let $a = \sup \lambda(B)$. Suppose $b \in B$; for every $c \in \lambda(B)$, $c \leqslant b$; hence b is an upper bound of $\lambda(B)$; thus $a \leqslant b$. This is true for each $b \in B$, so we conclude that *a is a lower bound of B*. Now if d is *any* lower bound of B, then

$d \in \lambda(B)$, so $a \geqslant d$ because a is an upper bound of $\lambda(B)$. We have proved that a is the greatest lower bound of B. ■

Dual If $\upsilon(B)$ has an inf in A, then B has a sup in A and $\sup B = \inf \upsilon(B)$.

4.33 Definition Let A be a partially ordered class. If every nonempty subclass of A that is bounded above has a sup, then A is said to be *conditionally complete.*

We have the following alternative definition of *conditionally complete*: A is called conditionally complete if every nonempty subclass of A that is bounded below has an inf. Our next theorem establishes the equivalence of the two definitions.

4.34 Theorem The following two conditions are equivalent:

i) Every nonempty subclass of A that is bounded above has a sup.
ii) Every nonempty subclass of A that is bounded below has an inf.

Proof

a) Suppose that (i) holds; let B be a nonempty subclass of A which is bounded below, that is, $\lambda(B) \neq \varnothing$. Each element of B is an upper bound of $\lambda(B)$, hence $\lambda(B)$ is bounded above; thus $\lambda(B)$ has a sup. But, by 4.32, it follows that B has an inf.

b) The converse is the dual of the result we have just proven. ■

EXERCISES 4.3

1. Suppose $B \subseteq A$; prove that if B has a greatest element b, then $b = \sup B$.
2. Suppose $B \subseteq A$; prove that $\upsilon(B) = \upsilon(\lambda(\upsilon(B)))$. [*Hint:* Use 4.29 and 4.31.]
3. Suppose $B \subseteq A$ and $C \subseteq A$; prove that $\lambda(B \cup C) = \lambda(B) \cap \lambda(C)$.
4. Suppose $B \subseteq A$; prove that if B has a sup b, then $\lambda(\upsilon(B)) \cap \upsilon(B) = \{b\}$.
5. Suppose $C \subseteq B$ and $B \subseteq A$; prove that $\sup_A C \leqslant \sup_B C$.
6. Let B and C be subclasses of a partially ordered class A. Prove that if $\sup B = \sup C$, then $\upsilon(B) = \upsilon(C)$.
7. Let A and B be partially ordered classes, and let $f: A \to B$ be a strictly increasing function. Prove that if b is a maximal element of B, then each element of $\overleftarrow{f}(b)$ is a maximal element of A.
8. Let A and B be partially ordered classes and let $f: A \to B$ be an increasing function. Prove that if a is the greatest element of A, then $f(a)$ is the greatest element of $\overrightarrow{f}(A)$.
9. Let A and B be partially ordered classes, and let $f: A \to B$ be an increasing function;

suppose $C \subseteq A$. Prove that if c is an upper bound of C, then $f(c)$ is an upper bound of $\overline{f}(C)$.

10. Let A and B be partially ordered classes, and let $f: A \to B$ be an isomorphism. Prove each of the following:
    a) a is a maximal element of A iff $f(a)$ is a maximal element of B.
    b) a is the greatest element of A iff $f(a)$ is the greatest element of B.
    c) Suppose $C \subseteq A$; x is an upper bound of C iff $f(x)$ is an upper bound of $\overline{f}(C)$.
    d) $b = \sup C$ iff $f(b) = \sup \overline{f}(C)$.

11. Let A be a partially ordered class. Prove the following:
    a) If every subclass of A has a sup and an inf, then A has a least element and a greatest element. [*Hint*: Use 4.27.]
    b) The following two statements are equivalent: Every subclass of A has a sup; every subclass of A has an inf.

12. Let A and B be partially ordered classes. Prove the following:
    a) Suppose $A \times B$ is ordered lexicographically: if (a, b) is a maximal element of $A \times B$, then a is a maximal element of A.
    b) Suppose $A \times B$ is ordered antilexicographically. If (a, b) is a maximal element of $A \times B$, then b is a maximal element of B.

4 LATTICES

4.35 Definition Let A be a partially ordered class. If every doubelton $\{x, y\}$ in A has a sup and an inf, then A is called a *lattice.*

When dealing with lattices it is customary to denote $\sup\{x, y\}$ by $x \vee y$ and $\inf\{x, y\}$ by $x \wedge y$. If A is a lattice, $x \vee y$ is often called the *join* of x and y, and $x \wedge y$ is often called the *meet* of x and y; the expression $x \vee y$ is read "x join y" and the expression $x \wedge y$ is read "x meet y."

Note the following simple consequences of our definition. If a and b are arbitrary elements of a lattice A, then

4.36 $a \leqslant a \vee b$ and $b \leqslant a \vee b$

because $a \vee b$ is an upper bound of a and b. Furthermore, if $c \in A$, then

4.37 $a \leqslant c$ and $b \leqslant c \Rightarrow a \vee b \leqslant c$;

in other words, if c is an upper bound of a and b, then $a \vee b \leqslant c$ because $a \vee b$ is the *least* upper bound of a and b.

For analogous reasons, we have

4.38 $a \wedge b \leqslant a$ and $a \wedge b \leqslant b$, and

4.39 $c \leqslant a$ and $c \leqslant b \Rightarrow c \leqslant a \wedge b$.

4.40 Theorem Let A be a lattice; the join and the meet have the following properties:

L1. $x \vee x = x$ and $x \wedge x = x$.

L2. $x \vee y = y \vee x$ and $x \wedge y = y \wedge x$.

L3. $(x \vee y) \vee z = x \vee (y \vee z)$ and $(x \wedge y) \wedge z = x \wedge (y \wedge z)$.

L4. $(x \vee y) \wedge x = x$ and $(x \wedge y) \vee x = x$.

Proof. L1 and L2 are immediate consequences of the definitions of sup and inf.

L3. First, we will prove that

$$(x \vee y) \vee z \leqslant x \vee (y \vee z);$$

by 4.36,

$$x \leqslant x \vee (y \vee z) \qquad \text{and} \qquad y \leqslant y \vee z \leqslant x \vee (y \vee z),$$

hence, by 4.37, $x \vee y \leqslant x \vee (y \vee z)$; furthermore, by 4.36,

$$z \leqslant y \vee z \leqslant x \vee (y \vee z),$$

so by 4.37, $(x \vee y) \vee z \leqslant x \vee (y \vee z)$. The inequality $x \vee (y \vee z) \leqslant (x \vee y) \vee z$ is proven in the same way, hence

$$(x \vee y) \vee z = x \vee (y \vee z).$$

The dual of this result is $(x \wedge y) \wedge z = x \wedge (y \wedge z)$.

L4. To prove that $(x \vee y) \wedge x = x$ is to prove that

$$x = \inf\{x \vee y, x\}.$$

Now x is a lower bound of $\{x \vee y, x\}$ because $x \leqslant x \vee y$ by 4.36 and obviously $x \leqslant x$. Furthermore, if z is any lower bound of $\{x \vee y, x\}$, then $z \leqslant x$; thus x is the greatest lower bound of $\{x \vee y, x\}$. This proves that $(x \vee y) \wedge x = x$; the dual of this result is $(x \wedge y) \vee x = x$. ■

A lattice may alternatively be defined as an algebraic system with two operations $\vee$ and $\wedge$ which have properties L1 through L4. This fact, which is of great importance in the study of lattices, is a consequence of the following theorem.

4.41 Theorem Let A be a class in which two operations denoted $\vee$ and $\wedge$ are given and have properties L1 through L4. We define a relation in A, to be denoted by the symbol $\leqslant$, as follows:

4.42 $x \leqslant y$ if and only if $x \vee y = y$.

Then $\leqslant$ is an order relation in A, and A is a lattice.

Proof. First, let us prove that the relation $\leqslant$ defined above is an order relation.

Reflexive. By L1, $x \vee x = x$; hence, by 4.42, $x \leqslant x$.

Antisymmetric. Suppose that $x \leqslant y$ and $y \leqslant x$; by 4.42, $x \vee y = y$ and $y \vee x = x$; but by L2, $x \vee y = y \vee x$, hence $x = y$.

Transitive. Suppose that $x \leqslant y$ and $y \leqslant z$; by 4.42, $x \vee y = y$ and $y \vee z = z$; thus

$$x \vee z = x \vee (y \vee z) = (x \vee y) \vee z = y \vee z = z;$$

so by 4.42, $x \leqslant z$.

Next, we will prove that $x \vee y$ is the least upper bound of x and y:

$$x \vee (x \vee y) = (x \vee x) \vee y = x \vee y,$$

so by 4.42, $x \leqslant x \vee y$; analogously, $y \leqslant x \vee y$, hence $x \vee y$ is an upper bound of x and y. Now if z is any upper bound of x and y, that is, $x \leqslant z$ and $y \leqslant z$, then $x \vee z = z$ and $y \vee z = z$; thus

$$(x \vee y) \vee z = x \vee (y \vee z) = x \vee z = z,$$

so by 4.42, $x \vee y \leqslant z$. This proves that $x \vee y = \sup\{x, y\}$. The proof that $x \wedge y = \inf\{x, y\}$ is left as an exercise for the reader. We conclude that A is a lattice. ■

It follows from 4.40 and 4.41 that a lattice may be defined in two distinct ways: as a partially ordered class in which every pair of elements has a sup and an inf or, alternatively, as an algebraic system with two operations satisfying rules L1 through L4.

4.43 Definition Let A be a lattice, and let B be a subclass of A. If

$$x \in B \text{ and } y \in B \Rightarrow x \vee y \in B \text{ and } x \wedge y \in B,$$

then B is called a *sublattice* of A.

4.44 Definition A *Boolean algebra* is defined to be a lattice A with the following additional properties:

L5. There is an element $0 \in A$ and an element $1 \in A$ such that for each $x \in A$, $x \vee 0 = x$ and $x \wedge 1 = x$.

L6. For each $x \in A$ there is an element $x' \in A$ such that

$$x \wedge x' = 0 \qquad \text{and} \qquad x \vee x' = 1.$$

L7. $x \vee (y \wedge z) = (x \vee y) \wedge (x \vee z)$ and $x \wedge (y \vee z) = (x \wedge y) \vee (x \wedge z)$.

The algebra of classes is an example of a Boolean algebra; indeed, by 1.22, 1.25, and 1.26, the operations $\cup$, $\cap$ and $'$ satisfy L1 through L7. We have noted independently (4.26) that $A \cup B = \sup\{A, B\}$ and $A \cap B = \inf\{A, B\}$. Another example of a Boolean algebra is the algebra of sentences, with the operations of conjunction, disjunction, and negation. In addition, Boolean algebra has a variety of applications in many areas of science and technology.

4.45 Definition Let A be a partially ordered class; A is called a *complete lattice* if every subclass of A has a sup. Alternatively, A is called a complete lattice if every subclass of A has an inf.

The purpose of our next theorem is to show that these two alternative definitions are equivalent. It will follow that if A is a complete lattice (in the sense of either definition), then every subclass of A has a sup and an inf; in particular, every doubleton in A has a sup and an inf, hence we are justified in calling A a lattice.

4.46 Theorem Let A be a partially ordered class; the following two conditions are equivalent:

i) every subclass of A has a sup;
ii) every subclass of A has an inf.

Proof. Let us assume that (i) holds; it follows that A has a sup, which is necessarily the greatest element of A, and $\varnothing$ has a sup, which is the least element of A (see 4.27). Let M designate the greatest element of A, and let m designate the least element of A. Let B be an arbitrary subclass of A. If $B = \varnothing$, then $\inf B = M$ (see 4.27); if $B \neq \varnothing$, then B is bounded below by m, hence by Theorem 4.34, B has an inf. Thus, (ii) holds; the converse is the dual of what we have just proven. ■

4.47 Example Let A be an arbitrary set and let $\mathscr{G}$ be the set of all the equivalence relations in A, ordered by inclusion. $\mathscr{G}$ has a least element, namely the relation I_A, and a greatest element, namely the relation $A \times A$. Furthermore, if $\{G_i\}_{i \in I}$ is any subset of $\mathscr{G}$, then $\bigcap_{i \in I} G_i$ is an element of $\mathscr{G}$ (see Exercise 4, Exercise Set 3.2); as we have seen (4.26), $\bigcap_{i \in I} G_i$ is the greatest lower bound of $\{G_i\}_{i \in I}$. Thus, $\mathscr{G}$ is a complete lattice by 4.46(ii).

EXERCISES 4.4

1. Let A be a class with two operations $\vee$ and $\wedge$ which satisfy L1 through L4. Prove that $x \vee y = y$ if and only if $x \wedge y = x$.
2. In Theorem 4.41, prove that $x \wedge y = \inf\{x, y\}$. [*Hint:* Use the result of Exercise 1.]
3. Let A be a lattice. Prove that the following statements are true.
   a) If $a \leqslant b$, then $\forall x \in A, a \vee x \leqslant b \vee x$ and $a \wedge x \leqslant b \wedge x$.
   b) If $a \leqslant b$ and $c \leqslant d$, then $a \vee c \leqslant b \vee d$ and $a \wedge c \leqslant b \wedge d$.
4. Let A be a lattice. If $[a, b]$ and $[c, d]$ are closed intervals of A, prove that
$$[a, b] \cap [c, d] = [a \vee c, b \wedge d].$$
(See Exercise 8, Exercise Set 4.2, for a definition of closed interval.)
5. Let A be a lattice; prove that every closed interval $[a, b]$ of A is a sublattice of A.
6. Let A be a lattice; if $a \in A$, let $I_a = \{x \in A \mid x \leqslant a\}$. Prove that I_a is a sublattice of A.
7. By a distributive lattice we mean a class with two operations satisfying L1 through L4 and L7. If A is a distributive lattice, prove that
$$c \vee x = c \vee y \text{ and } c \wedge x = c \wedge y \Rightarrow x = y.$$
8. In an arbitrary lattice A, prove the so-called "distributive inequalities"
$$x \wedge (y \vee z) \geqslant (x \wedge y) \vee (x \wedge z) \qquad \text{and} \qquad x \vee (y \wedge z) \leqslant (x \vee y) \wedge (x \vee z).$$
9. Let A be a lattice and let $x, y, z \in A$. Prove that if $x \leqslant z$, then
$$x \vee (y \wedge z) \leqslant (x \vee y) \wedge z.$$
10. Draw the line diagram of a lattice A and a subclass $B \subseteq A$ such that B is not a sublattice of A.
11. Draw the line diagram of two lattices whose intersection is not a lattice.
12. Draw the line diagram of a lattice which is not a Boolean algebra.
13. Let A be a partially ordered set; prove that the class of all the cuts of A (ordered by inclusion on the "left components" L) is a complete lattice.
14. Let A be a partially ordered set; prove that the class of all the convex subsets of A is a complete lattice.
15. Draw the line diagram of a partially ordered class which is conditionally complete, but is not a complete lattice.

5 FULLY ORDERED CLASSES. WELL-ORDERED CLASSES

Let A be a partially ordered class. As previously stated, A is said to be *fully ordered* if every two elements of A are comparable.

Examples of fully ordered classes are: the class $\mathbb{Z}$ of the integers, the class $\mathbb{Q}$ of the rational numbers, and the class $\mathbb{R}$ of the real numbers. One can easily see that every subclass of a fully ordered class is fully ordered. It is evident, too, that every fully ordered class is a lattice.

4.48 Theorem Let $f: A \to B$ be a function, where A is a fully ordered class and B is a partially ordered class. If $f: A \to B$ is bijective and increasing, it is an isomorphism.

Proof. Suppose $f(x) \leqslant f(y)$; since x and y are comparable, either $x \leqslant y$ or $y < x$. If $y < x$, then $f(y) \leqslant f(x)$; but $f(y) = f(x)$ would imply $y = x$, so we must have $f(y) < f(x)$. This is contrary to our assumption, hence $x \leqslant y$. ■

4.49 Definition Let A be a partially ordered class. A is said to be *well ordered* if every nonempty subclass of A has a least element.

If A is well ordered, then A is fully ordered; for if $x \in A$ and $y \in A$, then the doubleton $\{x, y\}$ has a least element, which is either x or y; hence $x \leqslant y$ or $y \leqslant x$.

If A is well ordered, then A is conditionally complete; for if B is a subclass of A and $\upsilon(B) \neq \varnothing$, then $\upsilon(B)$ has a least element which is by definition the sup of B.

4.50 Definition Let A be a partially ordered class, and suppose $a \in A$. An element $b \in A$ is called the *immediate successor* of a if $a < b$ and there is no element c in A such that $a < c < b$.

4.51 *Remark.* If A is a well-ordered class, then every element of A (with the exception of the greatest element of A, if it exists) has an immediate successor. Indeed, if $x \in A$ and x is not the greatest element of A, then the class $T = \{y \in A \mid y > x\}$ is nonempty, hence T has a least element which is obviously the immediate successor of x.

Let A be a nonempty well-ordered class. By 4.49, A has a least element, which may be denoted by x_1; if x_1 is not the greatest element (that is, the only element) of A, then by 4.51, x_1 has an immediate successor, which may be denoted by x_2; again, if x_2 is not the greatest element of A, then by 4.51, x_2 has an immediate successor, which may be denoted by x_3; and so on.

Examples

4.52 The class of numbers $\{1, 2, 3, 4, 5\}$, ordered in the usual way, is a well-ordered class.

4.53 The class $N = \{1, 2, 3, 4, \ldots\}$ of all the positive integers, ordered in the usual way, is a well-ordered class.

4.54 $W = \{0, \frac{1}{2}, \frac{3}{4}, \frac{7}{8}, \cdots, 1, 1 + \frac{1}{2}, 1 + \frac{3}{4}, 1 + \frac{7}{8}, \cdots, 2, 2 + \frac{1}{2}, 2 + \frac{3}{4}, 2 + \frac{7}{8}, \cdots\},$

ordered in the usual way, is a well-ordered subclass of the real numbers.

We may define y to be an *immediate predecessor* of x if and only if x is an immediate successor of y; note that in Examples 4.52 and 4.53, only the least element of the class (namely 1) does not have an immediate predecessor; however, in Example 4.54, there are three elements (namely 0, 1 and 2) which do not have an immediate predecessor.

Note that the class $\mathbb{Z}$ of all the integers, the class $\mathbb{Q}$ of the rational numbers, and the class $\mathbb{R}$ of the real numbers are *not* well ordered. Indeed,

$$V = \{\ldots, -3, -2, -1, 0\}$$

is a subclass of each of these classes, and V does not have a least element.

4.55 Definition Let A be a partially ordered class; we define a *section* of A to be a subclass $B \subseteq A$ with the following property:

$$\forall x \in A, \quad \text{if } y \in B \text{ and } x \leqslant y \qquad \text{then} \quad x \in B.$$

4.56 Theorem Let A be a well-ordered class; B is a section of A if and only if $B = A$ or B is an initial segment of A.

Proof

i) If $B = A$ or B is an initial segment of A, then obviously B is a section of A.

ii) Conversely, suppose B is a section of A; if $B = A$, we are done; thus, suppose $B \neq A$, that is, $A - B \neq \varnothing$. Because A is a well-ordered class, $A - B$ has a least element which we denote by m; we will show that $B = S_m$. Well,

$$x \in S_m \Rightarrow x < m \Rightarrow x \in B$$

(because m is the *least* element of $A - B$); conversely, suppose $x \in B$: if $m \leqslant x$, then $m \in B$ by 4.55, and this contradicts our choice of m; thus $x < m$, so $x \in S_m$. ■

One of the most important features of well-ordering is the fact that induction can be used to prove theorems about all the elements of a well-ordered class. This fact is given in the following theorem.

4.57 Theorem (*Principle of Transfinite Induction*). Let A be a well-ordered class, and let $P(x)$ be a statement which is either true or false for each element $x \in A$; suppose the following condition holds:

Ind. If $P(y)$ is true for every $y < x$, then $P(x)$ is true.

In that case, $P(x)$ is true for every element $x \in A$.

Proof. Suppose that $P(x)$ is *not* true for every $x \in A$; then the class $\{y \in A \mid P(y)$ is false$\}$ is nonempty; hence, by 4.49, has a least element m. Now $P(x)$ is true for every $x < m$, so by Ind, $P(m)$ is true; but we chose m to be the least element of $\{y \in A \mid P(y)$ is false$\}$, so $P(m)$ is false. This contradiction proves that $P(x)$ must be true for every $x \in A$. ∎

EXERCISES 4.5

1. Let A be a fully ordered set. Prove that the set of all sections of A (ordered by inclusion) is fully ordered.
2. Let A be a fully ordered class, let B be a partially ordered class, and let $f: A \to B$ be an increasing function. Prove that f is injective if and only if f is strictly increasing.
3. Let A be fully ordered class, let B be a partially ordered class, and let $f: A \to B$ be a bijective function. Prove that if f is increasing, then f is an isomorphism.
4. Let A be a fully ordered class and let $\{L, U\}$ be a partition of A. Prove that (L, U) is a cut of A if and only if $\forall x \in L$ and $\forall y \in U$, $x \leqslant y$.
5. Let A be a fully ordered class. Prove that if B and C are convex subclasses of A and $B \cap C \neq \varnothing$, then $B \cup C$ is convex.
6. Let A be a fully ordered class. Let B and C be convex subclasses of A and suppose $B \cap C \neq \varnothing$. Prove that every upper bound of $B \cap C$ is an upper bound of B or an upper bound of C. Conclude that
$$\upsilon(B \cap C) = \upsilon(B) \cup \upsilon(C).$$
7. Let A be a well-ordered class. If $x \in A$, prove that the immediate successor of x and the immediate predecessor of x (if it exists) are unique.
8. Let A be a partially ordered class; prove that B is a section of A if and only if $(B, A - B)$ is a cut of A.
9. Let A be a well-ordered class; prove the following:
   a) The intersection of any family of sections of A is a section of A.
   b) The union of any family of sections of A is a section of A.
10. Let A be a well-ordered class and let B and C be initial segments of A. Prove that if $B \subset C$, then B is an initial segment of C.
11. Let A be a fully ordered class. Let $B \subseteq A$ and $m \in B$; prove that B has a least element if and only if $S_m \cap B$ has a least element.
12. Let A be a fully ordered class. Prove that A is well ordered if and only if every initial segment of A is well ordered. [*Hint:* Use the result of Exercise 11, above.]
13. Let A be a well-ordered class. If $a \in A$, let a' designate the immediate successor of a, and let a'' designate the immediate predecessor of a (if it exists). Prove the following:
   a) $a \leqslant b$ if and only if $a' \leqslant b'$. b) $a = b$ if and only if $a' = b'$.

c) $a < b$ if and only if $a' < b'$. d) $a = b$ if and only if $a'' = b''$.
e) $a = b'$ if and only if $a'' = b$.

14. Let A be a well-ordered class; if $a \in A$, let a' designate the immediate successor of a. An element q in A will be called a *limit element* of A if q is not the least element of A and q does not have an immediate predecessor. Prove the following:
a) q is a limit element of A iff $[a < q \Rightarrow a' < q]$.
b) q is a limit element of A iff $q = \sup\{x \in A \mid x < q\}$.

6 ISOMORPHISM BETWEEN WELL-ORDERED CLASSES

The purpose of this section is to prove a remarkable property of well-ordered classes: if A and B are any two well-ordered classes, either A is isomorphic with B, or else one of the two is isomorphic with an initial segment of the other. What this means, roughly speaking, is that *well-ordered classes do not differ from one another except in their size.* This fact has many important applications in mathematics, and will be essential to our later discussion of infinite sets and cardinal and ordinal numbers. We begin by proving three preparatory lemmas.

4.58 Lemma Let A be a well-ordered class, and let f be an isomorphism from A to a subclass of A. Then $x \leqslant f(x)$, $\forall x \in A$.

Proof. Assume, on the contrary, that the class $P = \{x \in A \mid x > f(x)\}$ is non-empty, and let a be the least element of P; hence, in particular, $f(a) < a$. We now have

$$f(f(a)) < f(a) < a,$$

so $f(a) \in P$, which is impossible because a is the least element of P. Thus $P = \varnothing$, and the lemma is proved. ■

4.59 Lemma Let A be a well-ordered class. There is no isomorphism from A to a subclass of an initial segment of A.

Proof. Assume, on the contrary, that f is an isomorphism from A to a subclass of an initial segment S_a of A. By Lemma 4.58, $a \leqslant f(a)$, so $f(a) \notin S_a$; this is impossible, for the range of f is assumed to be a subclass of S_a. Hence a function f of the kind we assumed cannot exist. ■

4.60 Corollary No well-ordered class is isomorphic with an initial segment of itself.

4.61 Lemma Let A and B be well-ordered classes. If A is isomorphic with an initial segment of B, then B is not isomorphic with any subclass of A.

Proof. Let $f: A \to S_b$ be an isomorphism from A to an initial segment of B. Assume there exists an isomorphism $g: B \to C$ where $C \subseteq A$. Obviously $g: B \to A$ is a function; $g: B \to A$ and $f: A \to S_b$ are both injective and increasing, hence their composite $f \circ g: B \to S_b$ is injective and increasing; that is, by 4.48, $f \circ g$ is an isomorphism from B to its range which is a subclass of S_b. However, by Lemma 4.59, this is impossible; hence the isomorphism g that was assumed cannot exist. ■

4.62 Theorem Let A and B be well-ordered classes; exactly one of the following three cases must hold:

i) A is isomorphic with B.
ii) A is isomorphic with an initial segment of B.
iii) B is isomorphic with an initial segment of A.

Proof. We begin by proving that the following holds in any well-ordered class X.

I. Let S_x and S_y be initial segments of X; if $x < y$, then S_x is an initial segment of S_y.

Indeed, if $x < y$, then clearly $S_x \subset S_y$; furthermore, S_x is a section of S_y, for

$$[u \in S_x \text{ and } v \leqslant u] \Rightarrow v \leqslant u < x \Rightarrow v \in S_x;$$

thus by 4.56 (note that $S_x \neq S_y$ because $x \neq y$) we conclude that S_x is an initial segment of S_y.

Now let A and B be well-ordered classes, and let C be the following subclass of A:

$$C = \{x \in A \mid \exists r \in B \ni S_x \approxeq S_r\}.$$

If $x \in C$, there is no more than one $r \in B$ such that $S_x \approxeq S_r$; for suppose $S_x \approxeq S_r$ and $S_x \approxeq S_t$, where $r \neq t$, say $r < t$. By I, S_r is an initial segment of S_t; but $S_r \approxeq S_x \approxeq S_t$, and this is impossible by 4.60; thus for each $x \in C$, the element $r \in B$ such that $S_x \approxeq S_r$ is unique. Let us designate the unique $r \in B$ corresponding to x by $F(x)$; thus $F: C \to B$ is a function. In particular, if $D = \operatorname{ran} F$, then $F: C \to D$ is a function; we will show next that $F: C \to D$ is an isomorphism.

F is injective. Suppose $F(u) = F(v) = r$, that is, $S_u \approxeq S_r \approxeq S_v$. If $u \neq v$, say $u < v$, then by I, S_u is an initial segment of S_v, and this is impossible by 4.60. We conclude that $u = v$.

F is increasing. Suppose $u \leqslant v$, where $F(u) = r$ and $F(v) = t$; hence $S_u \approxeq S_r$ and $S_v \approxeq S_t$. Assume that $t < r$, hence by I, S_t is an initial segment of S_r; now $S_u \subseteq S_v$, so

a) S_v is isomorphic with an initial segment of S_r, and
b) S_r is isomorphic with a subclass of S_v.

This is impossible by 4.61, hence we conclude that $r \leqslant t$, that is, $F(u) \leqslant F(v)$. It follows, by 4.48, that $F: C \to D$ is an isomorphism.

Next, we will show that C is a section of A; that is, given $c \in C$ and $x < c$, we will prove that $x \in C$. If $F(c) = r$, then $S_c \approx S_r$, that is, there exists an isomorphism $g: S_c \to S_r$. It is a simple exercise to prove that

$$g_{[S_x]}: S_x \to S_{g(x)}$$

is an isomorphism; the details are left as an exercise for the reader. Thus $S_x \approx S_{g(x)}$, so $x \in C$.

An analogous argument shows that D is a section of B. Thus by 4.56, our theorem will be proven if we can show that the following is *false*:

$$C \text{ is an initial segment of } A, \text{ and } D \text{ is an initial segment of } B.$$

Indeed, suppose the above to be true: say $C = S_x$ and $D = S_r$; we have proven that $F: C \to D$ is an isomorphism, that is, $C \approx D$, so $S_x \approx S_r$. But then $x \in C$, that is, $x \in S_x$, which is absurd; this proves that one of the conditions (i), (ii) or (iii) necessarily holds. The fact that no two of these conditions holds simultaneously follows from 4.60 and 4.61. ■

4.63 Corollary Let A be a well-ordered class; every subclass of A is isomorphic with A or an initial segment of A.

Proof. If B is a subclass of A, then B is well ordered; hence by 4.62, $B \approx A$, or B is isomorphic with an initial segment of A, or A is isomorphic with an initial segment of B. In order to prove our result we must show that the last case cannot hold; indeed, suppose it does: then by 4.61, B is not isomorphic with any subclass of A. But $B \approx B$ and B is a subclass of A, so we have a contradiction; thus the last case cannot hold. ■

EXERCISES 4.6

1. In the proof of Theorem 4.62, prove that $g_{[S_x]}: S_x \to S_{g(x)}$ is an isomorphism.
2. In the proof of Theorem 4.62, prove that D is a section of B.
3. Let A be a well-ordered class. Prove that the identity mapping I_A is the only isomorphism from A to A.
4. Let A and B be well-ordered classes. Prove that if $f: A \to B$ and $g: B \to A$ are isomorphisms, then $g = f^{-1}$.
5. Let A and B be well-ordered classes. Prove that there exists at most one isomorphism $f: A \to B$.
6. Let A and B be well-ordered classes. Prove that if A is isomorphic with a subclass of B, and B is isomorphic with a subclass of A, then A is isomorphic with B.

7. Let A and B be well-ordered classes. Prove that if A is isomorphic with a class containing B, and B is isomorphic with a class containing A, then A is isomorphic with B.

8. Let A and B be well-ordered classes. Suppose that A has no greatest element; suppose that every element of B (except the least element) has an immediate predecessor. Prove that B is isomorphic with a section of A.

5

The Axiom of Choice and Related Principles

1 INTRODUCTION

From here on, throughout the remainder of this book, we will be concerned mainly with sets. We explained in Section 7 of Chapter 1 that while the notion of *class* is appealing because of its intuitive simplicity and generality, it suffers from a serious deficiency: given an arbitrary property of classes, we cannot form the class of all classes having that property. Sets do not suffer from this deficiency; we have already proven (1.49) that if we are given some property of sets, we may form the class of all the sets which have that property. In particular, we have shown (1.52) that if A is a set, then the class of all the subsets of A which satisfy any given property is again a set. We shall need to use these facts frequently in the following pages.

In this chapter we will discuss a concept which is one of the most important, and at the same time one of the most controversial, principles of mathematics. In 1904, Zermelo brought attention to an assumption which is used implicitly in a variety of mathematical arguments. This assumption does not follow from any previously known postulates of mathematics or logic, hence it must be taken as a new axiom; Zermelo called it the *Axiom of Choice.* The Axiom of Choice has significant implications in many branches of mathematics, and consequences so powerful as, sometimes, to defy credibility. The controversy over this principle continues in our day; we will present some of its aspects in this chapter.

In order to illustrate where the Axiom of Choice intrudes in common mathematical arguments, let us examine the following statement.

5.1 Let A be a nonempty, partially ordered set, and suppose that there are no maximal elements in A; then there exists a nonterminating, increasing sequence $x_1 < x_2 < x_3 < \cdots$ of elements of A.

Proof. A is nonempty by hypothesis, hence we may select an arbitrary element of A and call it x_1. By induction, suppose that we are given $x_1 < x_2 < \cdots < x_n$; we define A_n to be the set of all the elements $x \in A$ such that $x > x_n$. A_n is nonempty, for if it were empty, then x_n would be maximal, contradicting one of our assumptions. We select an arbitrary element of A_n and call it x_{n+1}; thus we have $x_1 < \cdots < x_n < x_{n+1}$.

This inductive process defines an increasing sequence $S_n = \{x_1, x_2, \ldots, x_n\}$ for each natural number n; that is, it gives us

$$S_1 = \{x_1\}, \quad S_2 = \{x_1, x_2\}, \quad S_3 = \{x_1, x_2, x_3\},$$

and so on. Now if we let $S = \bigcup_{n \in N} S_n$ (where N is the set of all the positive integers), then S is the nonterminating sequence $x_1 < x_2 < x_3 < \cdots$ that we are seeking.

A careful examination of the above argument will reveal that we have used an assumption which is by no means self-evident or undisputably plausible. What we have, in fact, done is to assume that we can make an *infinite succession of arbitrary choices.* It is common enough, in mathematics, to make *one* arbitrary choice (we do this every time we say "let x be an arbitrary element of A"), and experience confirms that we can make a finite succession of choices; but to make an infinite succession of choices is to carry an argument through an infinite number of steps—and nothing in our experience or in the logic we habitually use justifies such a process.

In the proof of 5.1, it was necessary to choose the elements x_1, x_2, x_3, etc., *in succession,* for each choice depended on the preceding ones. The fact that the choices are successive may appear to be the most disturbing element in the whole proof, for this involves a time factor (an infinite *succession* of acts, each one requiring a certain amount of time, would take infinitely long). However, the argument can be altered in such a way that all the choices are made simultaneously and independently of one another; we proceed as follows.

Let us admit that from each nonempty subset $B \subseteq A$ it is possible to choose an arbitrary element $\mathbf{r}_B$, to be called the "representative" of B. Note that in this case each choice is independent of the others; hence, in a manner of speaking, all the choices can be made simultaneously. Returning to the proof of 5.1, if x_n and A_n are given, we may define x_{n+1} to be the representative of A_n; in other words, instead of choosing representatives for A_1, A_2, A_3, etc., in succession, we have chosen representatives for *all* the nonempty subsets of A in advance. (Of course, this requires that we make many more choices than are needed for our original argument, but this is the price we must pay to substitute simultaneous choices for successive ones.)

The preceding paragraph makes it clear that the *successive* nature of the choices is not the crux of the problem; the problem is: *Can we make infinitely many choices*—be they successive or simultaneous?

It is worth noting that in certain particular cases the answer to this question is an obvious "yes." For example, if A is a well-ordered set, we may define the "representative" of each nonempty subset $B \subseteq A$ to be the *least* element of B; because A is well ordered, we have a law at our disposal which *provides us* with a representative for each nonempty subset of A. The situation in 5.1, however, is completely different, for *we do not have any ready-made rule* which is able to furnish us with representatives. It is only the latter case—as in 5.1—which is of interest to us here.

In the proof of 5.1 we speak of "selecting" an element of A_n; clearly, we do not wish to introduce the notion of "selecting" as a new undefined concept of set theory, so we avoid the use of this word by letting an appropriate function "select" representatives.

5.2 Definition Let A be a set; let us agree to write $\mathscr{P}'(A)$ for $\mathscr{P}(A) - \{\varnothing\}$. By a *choice function* for A we mean a function $\mathbf{r}: \mathscr{P}'(A) \to A$ such that

$$\forall B \in \mathscr{P}'(A), \quad \mathbf{r}(B) \in B.$$

We will sometimes write $\mathbf{r}_B$ for $\mathbf{r}(B)$ and call $\mathbf{r}_B$ the *representative* of B.

5.3 Example Let $A = \{a, b, c\}$; an example of a choice function for A is the function $\mathbf{r}$ given in the following table.

B	$\mathbf{r}(B)$
$\{a, b, c\}$	a
$\{a, b\}$	a
$\{a, c\}$	c
$\{b, c\}$	c
$\{a\}$	a
$\{b\}$	b
$\{c\}$	c

In the light of Definition 5.2, the question we have been asking can be expressed as follows: **if A is a set, does there exist a choice function for A?** A crucial comment needs to be made at this point: the proof of 5.1 does not require that we *construct* a choice function, it requires merely that a choice function *exist*! Indeed, if $\mathbf{r}$ is a choice function for A, then—in the controversial step of the proof—we let $x_{n+1} = \mathbf{r}(A_n)$; if we are assured that $\mathbf{r}$ *exists*, there is no further difficulty.

The Axiom of Choice asserts that every set has a choice function; its intent is to *state the existence* of a choice function for every set, even where, admittedly, none can be actually constructed. It must be emphasized here that to state the existence of a choice function is quite a different thing from producing one, or even claiming that one can be produced. For we are merely asserting that among *all possible* functions from $\mathscr{P}'(A)$ to A, there is *one at least* which maps every B onto an element $x \in B$.

The essence of the Axiom of Choice is that it is an *existential* statement rather than a *constructive* one. Once again, it states that among all possible functions from $\mathscr{P}'(A)$ to A, there is at least one which satisfies the condition of Definition 5.2; this does not seem grossly unreasonable. The Axiom of Choice makes no claim that a choice function can be constructed; hence, it does not assert that the sequence of choices described in the proof of 5.1 can be effectively carried out—and indeed this is not necessary in order for the proof to work.

Before the reader decides whether or not the Axiom of Choice seems plausible to him, he should examine some of the equivalent propositions which are developed in the next few sections of this chapter. Some of them are very powerful indeed, and a rejection of any one of them would entail a rejection of all of them (including, of course, the Axiom of Choice). It is important to note that what all of these principles have in common is the fact that they are nonconstructive: they assert the existence of mathematical objects which cannot be explicitly produced. It is precisely *this* aspect of the Axiom of Choice and related principles which makes them unacceptable to the intuitionists (see page 17, Section 5, Chapter 0), who claim that mathematical existence and constructibility are one.

Without reentering into the arguments for and against admitting nonconstructive propositions into mathematics, we can say this: intuitively, the Axiom of Choice cannot be rejected outright, nor can we feel truly certain of its validity. A more important consideration, however, is the fact that it has been proven that the Axiom of Choice does not contradict the other axioms of set theory, nor is it a consequence of them. Thus it has the same status as another famous axiom in mathematics, namely Euclid's "Fifth Postulate." We can have a "standard" set theory in which we postulate the Axiom of Choice, and "nonstandard" set theories in which we postulate alternatives to the Axiom of Choice.

In conclusion, since the Axiom of Choice is neither a consequence of the other axioms of set theory nor in conflict with them, it is impossible to make a decision *pro* or *con* on purely logical grounds. Since the Axiom of Choice involves an area of mathematics (namely, infinite sets) which is outside the realm of our experience, it will never be possible to confirm it or deny it by "observation." In the final analysis, the decision must be a purely personal one for each individual mathematician to make; it is a matter of personal taste.

2 THE AXIOM OF CHOICE

In the preceding section, we have given the background for our next axiom:

A8 (*Axiom of Choice*). Every set has a choice function.

In the literature there are several other ways of stating the Axiom of

Choice which are equivalent to our Axiom A8. We shall present two of these statements here, and several more in the exercises following this section.

Ch 1 Let $\mathscr{A}$ be a set whose elements are mutually disjoint, nonempty sets. There exists a set C which consists of exactly one element from each $A \in \mathscr{A}$.

If $\mathscr{A}$ is a family of disjoint, nonempty sets, then the set C described in Ch 1 is called a *choice set* for $\mathscr{A}$. Thus, Ch 1 asserts that every set of disjoint, nonempty sets has a choice set.

Ch 2 Let $\{A_i\}_{i \in I}$ be a set of sets. If I is nonempty and each A_i is nonempty, then $\prod_{i \in I} A_i$ is nonempty.

Let us show, first, that A8 $\Rightarrow$ Ch 1. Suppose $\mathscr{A}$ is a set whose elements are mutually disjoint, nonempty sets, and let

$$A = \bigcup_{X \in \mathscr{A}} X.$$

Clearly, $\mathscr{A} \subseteq \mathscr{P}'(A)$; by A8, there is a function $\mathbf{r} : \mathscr{P}'(A) \to A$ such that $\mathbf{r}(B) \in B$ for each $B \in \mathscr{P}'(A)$; if $C = \bar{\mathbf{r}}(\mathscr{A})$, it follows immediately that C is the set required in Ch. 1.

Next, Ch 1 $\Rightarrow$ A8. If A is a set and $B \subseteq A$, let $Q_B = \{(B, x) \mid x \in B\}$. If B and D are distinct, then Q_B and Q_D are disjoint, for Q_B consists of pairs (B, x), whereas Q_D consists of pairs (D, x). Thus the family $\{Q_B\}_{B \in \mathscr{P}'(A)}$ is a set of disjoint, nonempty sets*; it follows by Ch 1 that there exists a set C which consists of exactly one element (B, x) from each Q_B; it is easily verified that C is choice function for A.

The fact that A8 $\Rightarrow$ Ch 2 follows easily from the definition of a product of sets. Indeed, let $\{A_i\}_{i \in I}$ be a set of nonempty sets, and let

$$A = \bigcup_{i \in I} A_i;$$

by A8, there exists a function $\mathbf{r} : \mathscr{P}'(A) \to A$ such that $\mathbf{r}(B) \in B$ for each $B \in \mathscr{P}'(A)$; hence, in particular, $\mathbf{r}(A_i) \in A_i$ for each $i \in I$. If we define $\mathbf{a}$ by $\mathbf{a}(i) = \mathbf{r}(A_i)$, then $\mathbf{a}$ is a function from I to A such that $\mathbf{a}(i) \in A_i$ for each $i \in I$; that is, $\mathbf{a} \in \prod_{i \in I} A_i$. Thus $\prod_{i \in I} A_i$ is nonempty.

The fact that Ch 2 $\Rightarrow$ A8 can be proven by an argument similar to the above; the details are left to the reader.

---

* Note that for each $B \in \mathscr{P}'(A)$, $Q_B \subseteq \mathscr{P}(A) \times A$; thus

$$\{Q_B\}_{B \in \mathscr{P}'(A)} \subseteq \mathscr{P}[\mathscr{P}(A) \times A];$$

hence by A 3, A6, and 1.53, $\{Q_B\}_{B \in \mathscr{P}'(A)}$ is a set.

In Chapter 2 we promised to give a characterization of surjective functions whose proof depends on the Axiom of Choice.

5.4 Theorem Let A be a set and let $f: A \to B$ be a function; $f: A \to B$ is surjective if and only if there exists a function $g: B \to A \ni f \circ g = I_B$.

Proof

i) Suppose there exists a function $g: B \to A$ such that $f \circ g = I_B$; the proof that $f: A \to B$ is surjective is given in part (ii) of the proof of 2.24.

ii) Conversely, suppose that $f: A \to B$ is surjective; for each $y \in B, \overline{f}(y)$ is a nonempty subset of A. If $\mathbf{r}$ is a choice function for A, we define $g: B \to A$ by $g(y) = \mathbf{r}[\overline{f}(y)]$, $\forall y \in B$. In simple terms, for each $y \in B$ we let $g(y)$ be an arbitrary element of $\overline{f}(y)$. It is obvious that if $x = g(y)$, then $x \in \overline{f}(y)$, hence $f(x) = y$; thus

$$[f \circ g](y) = f(x) = y = I_B(y). \blacksquare$$

For the sake of simplicity we have proven Theorem 5.4 in the case where A is a set; using a slightly stronger form of the Axiom of Choice we can prove 5.4 in the more general case where A is any class; we omit the details.

EXERCISES 5.2

1. Let A be a set and let $f: A \to B$ be a surjective function. Prove that there exists a subset $C \subseteq A$ such that C is in one-to-one correspondence with B. [*Hint:* Use 5.4.]
2. Let A be a set, let $f: B \to C$ and $g: A \to C$ be functions, and suppose that ran $f \subseteq$ ran g. Prove that there exists a function $h: B \to A$ such that $g \circ h = f$. [*Hint:* Use the Axiom of Choice.]
3. Let $\{A_i\}_{i \in I}$ be an indexed family of classes, where I is a set. Prove that there exists $J \subseteq I$ such that
$$\{A_i \mid i \in I\} = \{A_j \mid j \in J\}$$
and, in $\{A_j\}_{j \in J}$, each A_j is indexed only once (that is, $A_i = A_j \Rightarrow i = j$). [*Hint:* Use Remark 2.38 and the Axiom of Choice.]
4. Prove that the statement of Theorem 5.4 implies the Axiom of Choice.

In each of the following problems a proposition is stated. Prove that this proposition is equivalent to the Axiom of Choice.

5. Let $\mathscr{A}$ be a set of disjoint, nonempty sets. There exists a function f, whose domain is $\mathscr{A}$, such that for all $A \in \mathscr{A}, f(A) \in A$.
6. Let E be a set and suppose $G \subseteq E \times E$. Let $A =$ dom G and $B =$ ran G; then there exists a function $f: A \to B$ such that $f \subseteq G$.

7. Let $\mathscr{A}$ be a set whose elements are nonempty sets, and let $A = \bigcup_{X \in \mathscr{A}} X$. Then, corresponding to every function $g: \mathscr{A} \to \mathscr{A}$, there exists a function $g^*: \mathscr{A} \to A$ such that $g^*(B) \in g(B)$.
8. Let B be a set and let $f: A \to B$ be a function; then there exist subsets $C \subseteq A$ and $g \subseteq f$ such that $g: C \to B$ is an injective function and $\text{ran } g = \text{ran } f$.

3 AN APPLICATION OF THE AXIOM OF CHOICE

The purpose of this section is to develop a consequence of the Axiom of Choice. The result we are about to prove is valuable as a stepping stone which will enable us to prove the important maximal principles that follow in the next section.

Let A be a partially ordered set such that every chain of A has a sup in A; assume that A has a least element p. We intend to show that there exists an element $a \in A$ such that a has no immediate successor.

In order to show this, we will suppose that every element $x \in A$ has an immediate successor; this assumption will lead to a contradiction.

If every element of A has an immediate successor, then we can define a function $f: A \to A$ such that for each $x \in A$, $f(x)$ is an immediate successor of x. Indeed, let T_x be the set of all the immediate successors of x; by the Axiom of Choice, there exists a choice function g such that $g(T_x) \in T_x$. We define f by letting $f(x) = g(T_x)$; clearly, $f(x)$ is an immediate successor of x.

5.5 Definition A subset $B \subseteq A$ is called a *p-sequence* if the following conditions are satisfied.

α) $p \in B$,
β) if $x \in B$, then $f(x) \in B$,
γ) if C is a chain of B, then $\sup C \in B$.

There *are* p-sequences; for example, A is a p-sequence.

5.6 Lemma Any intersection of p-sequences is a p-sequence.
The proof is left as an exercise for the reader.

Let P be the intersection of all the p-sequences. (Note that $P \neq \varnothing$ because $p \in P$). By 5.6, P is a p-sequence.

5.7 Definition An element $x \in P$ is called *select* if it is comparable with every element $y \in P$.

5.8 Lemma Suppose x is select, $y \in P$, and $y < x$. Then $f(y) \leq x$.

Proof. $y \in P$, P is a p-sequence, hence by (β), $f(y) \in P$. Now, x is select, so either $f(y) \leq x$ or $x < f(y)$. By hypothesis $y < x$; so if $x < f(y)$, we have $y < x < f(y)$, which contradicts the assertion that $f(y)$ is the immediate successor of y. Hence $f(y) \leq x$. ■

5.9 Lemma Suppose x is select. Let

$$B_x = \{y \in P \mid y \leq x \text{ or } y \geq f(x)\}.$$

Then B_x is a p-sequence.

Proof. We will show that B_x satisfies the three conditions which define a p-sequence.

α) Since p is the least element of A, $p \leq x$, hence $p \in B_x$.

β) Suppose $y \in B_x$; then $y \leq x$ or $y \geq f(x)$. Consider three cases:
1) $y < x$. Then $f(y) \leq x$ by 5.8, hence $f(y) \in B_x$.
2) $y = x$. Then $f(y) = f(x)$, thus $f(y) \geq f(x)$; hence $f(y) \in B_x$.
3) $y \geqslant f(x)$. But $f(y) > y$, so $f(y) > f(x)$; hence $f(y) \in B_x$.

In each case we conclude that $f(y) \in B_x$.

γ) If C is a chain of B_x, let $m = \sup C$. For each $y \in B_x$, $y \leq x$ or $y \geq f(x)$. If $\exists y \in C \ni y \geq f(x)$, then (since $m \geq y$) $m \geq f(x)$, so $m \in B_x$. Otherwise, $\forall y \in C$, $y \leq x$; thus x is an upper bound of C, so $m \leq x$. Thus again $m \in B_x$. ■

5.10 Corollary If x is select, then $\forall y \in P$, $y \leq x$ or $y \geq f(x)$.

Proof. B_x is a p-sequence; P is the intersection of all p-sequences; hence $P \subseteq B_x$. But $B_x \subseteq P$ by definition, hence $P = B_x$. So $\forall y \in P$, $y \leq x$ or $y \geq f(x)$. ■

5.11 Lemma The set of all select elements is a p-sequence.

Proof

α) p is select because it is less than (hence comparable to) each $y \in P$.

β) Suppose x is select, by 5.10, $\forall y \in P$, either $y \leq x$ (in which case $y \leq f(x)$ because $x < f(x)$) or $y \geq f(x)$. Thus $f(x)$ is select.

γ) Let C be a chain of select elements and let $m = \sup C$; let $y \in P$. If $\exists x \in C \ni y \leq x$, then $y \leq m$ (because $x \leqslant m$). Otherwise, $\forall x \in C$, $x \leq y$, hence y is an upper bound of C, so $m \leq y$. Thus m is select. ■

5.12 Corollary P is fully ordered.

Proof. The set S of all the select elements is a p-sequence; P is the intersection of all the p-sequences; hence $P \subseteq S$. But $S \subseteq P$ (by definition a select element is in P), so $P = S$. Thus each element of P is select, that is, is comparable to each element of P. ■

Corollary 5.12 produces a contradiction. Indeed, let $m = \sup P$; by condition (γ), $m \in P$ because P is a chain of P. But by condition (β), $f(m) \in P$, hence $f(m) \leq m$; this contradicts the assertion that $f(m)$ is an immediate successor of m. We conclude:

5.13 Theorem Let A be a partially ordered set such that (1) A has a least element p and (2) every chain of A has a sup in A. Then there is an element $x \in A$ which has no immediate successor.

4 MAXIMAL PRINCIPLES

The propositions we are about to develop are widely used in mathematics to prove theorems by "nonconstructive" methods. They assert the existence of mathematical objects which cannot be constructed. For example, in linear algebra a maximal principle can be used to prove that every vector space has a basis, although, in general, it is impossible to exhibit such a basis.

The maximal principles are consequences of the Axiom of Choice. Furthermore, as we shall verify in Section 6, they are equivalent to the Axiom of Choice.

5.14 Theorem (*Hausdorff's Maximal Principle*). Every partially ordered set has a maximal chain.

Proof. Let A be a partially ordered set, and let $\mathscr{S}$ be the set of all the chains of A, ordered by inclusion. $\mathscr{S}$ has a least element, namely the empty set. Now let $\mathscr{C}$ be a chain of $\mathscr{S}$ and let

$$K = \bigcup_{C \in \mathscr{C}} C;$$

we will show that $K \in \mathscr{S}$. Indeed, if $x, y \in K$, then $x \in D$ and $y \in E$ for some elements $D \in \mathscr{C}$ and $E \in \mathscr{C}$; but $\mathscr{C}$ is a chain of $\mathscr{S}$, hence $C \subseteq D$ or $D \subseteq C$, say $C \subseteq D$; thus $x, y \in D$. But D is a chain of A (remember that $\mathscr{S}$ is the set of all the chains of A), so x and y are comparable; this proves that K is a chain of A, that is, $K \in \mathscr{S}$. By 4.26, $K = \sup \mathscr{C}$; it follows that the conditions of Theorem 5.13 are satisfied by $\mathscr{S}$. Thus, by 5.13, there is an element $C \in \mathscr{S}$ which has no immediate successor; that is, there exists no $x \in A - C$ such that $C \cup \{x\}$ is a chain of A. Thus, clearly, C is a maximal chain. ■

5.15 Definition A partially ordered set A is said to be *inductive* if every chain of A has an upper bound in A.

5.16 Theorem (*Zorn's Lemma*). Every inductive set has at least one maximal element.

Proof. Let A be an inductive set; by 5.14, A has a maximal chain C; by 5.15, C has an upper bound m. Now suppose there exists an element $x \in A \ni x > m$; then $x \notin C$, but x is comparable with (to be exact, x is greater than) every element of C. Thus, $C \cup \{x\}$ is a chain, contradicting the assertion that C is a maximal chain; hence there exists no element $x \in A$ such that $x > m$, so m is a maximal element of A. ■

Theorems 5.14 and 5.16 can be stated in a somewhat stronger form as follows:

5.17 Theorem Every partially ordered set has a maximal well-ordered subset.

5.18 Theorem (Let us call a partially ordered set A *weakly inductive* if every well-ordered subset of A has an upper bound in A.) Every weakly inductive set has at least one maximal element.

The proofs of the last two theorems are similar to those of Theorems 5.14 and 5.16; they are left as an exercise for the reader.

EXERCISES 5.4

1. Let A be a partially ordered set and let $\mathscr{A}$ be the set of all the well-ordered subsets of A. For $C \in \mathscr{A}$ and $D \in \mathscr{A}$, define $C \dashv D$ if and only if C is a section of D.
   a) Prove that $\dashv$ is a partial order relation in $\mathscr{A}$.
   b) Prove that $\mathscr{A}$, ordered by $\dashv$, is inductive.
   c) Using part (b) and Zorn's Lemma, prove Theorem 5.17.
2. Use the result of Exercise 1, above, to prove Theorem 5.18.
3. Derive Hausdorff's Maximal Principle from Zorn's Lemma.
4. Prove that Zorn's Lemma is equivalent to the following: Let A be an inductive set and let $a \in A$; then A has at least one maximal element b such that $b \geqslant a$.
5. Prove that Hausdorff's Maximal Principle is equivalent to the following: If A is a partially ordered set and B is a chain of A, then A has a maximal chain C such that $B \subseteq C$.
6. Let A be any set with more than one element. Prove that there exists a bijective function $f: A \to A$ such that $f(x) \neq x, \forall x \in A$.
7. A set of sets $\mathscr{A}$ is said to be disjointed if $\forall C, D \in \mathscr{A}, C \cap D = \varnothing$. Let $\mathscr{F}$ be a set of sets; prove that $\mathscr{F}$ has a maximal disjointed subset.
8. Let A be a set and let $\mathscr{A}$ be a set of subsets of A; let $\mathscr{A}$ have the following property:

$B \in \mathscr{A}$ iff every finite subset of B belongs to $\mathscr{A}$; then $\mathscr{A}$ is said to be of *finite character.* Let $\mathscr{A}$ be ordered by inclusion and suppose $\mathscr{A}$ is of finite character.

a) Prove that $\mathscr{A}$ is an inductive set.
b) Prove that $\mathscr{A}$ has a maximal element.

9. Prove that every vector space V has a basis. [*Hint:* Consider the set $\mathscr{A}$ of all the linearly independent subsets of V. Use Zorn's Lemma; it is easily verified that any maximal linearly independent subset of V is a basis of V.]
10. Let G be a group and let A be an arbitrary subset of G such that A includes the identity element of G. Prove that among the subgroups of G which are subsets of A, there is a maximal one.

5 THE WELL-ORDERING THEOREM

The well-ordering theorem, which will be presented in this section, is one of the most important consequences of the Axiom of Choice and is an outstanding example of a nonconstructive proposition. It asserts that any set can be well ordered; that is, if A is any set, there exists an order relation G such that A, ordered by G, is a well-ordered set. The proof of the well-ordering theorem gives no indication how such a well-ordering of the elements of A is to be accomplished; it asserts merely that a well-ordering exists.

A finite set A can obviously be well ordered; for example, if $A = \{a, b, c\}$, then $a < b < c$, $b < a < c$, $c < b < a$ are different well-orderings of A. However, no method has yet been discovered to well-order sets such as the set $\mathbb{R}$ of the real numbers; in fact, in the opinion of most mathematicians, it is impossible to construct a well-ordering of $\mathbb{R}$. (This would mean giving a rule for rearranging the elements of $\mathbb{R}$ in such a way that $\mathbb{R}$ would then become a well-ordered set.)

Once again, the well-ordering theorem does *not* assert that every set can be *effectively* well ordered; it merely states that, among all possible graphs $G \subseteq A \times A$, there exists one at least which is an order relation which well-orders A.

Let us now prove the well-ordering theorem; note that the proof relies heavily on the Axiom of Choice.

Let A be an arbitrary set. We will consider pairs (B, G), where B is a subset of A, and G is an order relation in B which well-orders B.

Let $\mathscr{A}$ be the family of all such pairs (B, G). We introduce the symbol $\prec$ and define $(B, G) \prec (B', G')$ if and only if

5.19 a) $B \subseteq B'$, b) $G \subseteq G'$, c) $x \in B$ and $y \in B' - B \Rightarrow (x, y) \in G'$.

(Note that the last condition asserts, roughly, that all the elements of B precede all the elements of $B' - B$.) It is easy to verify that $\prec$ is an order relation in $\mathscr{A}$; the details are left as an exercise for the reader.

5.20 Lemma Let

$$\mathscr{C} = \{(B_i, G_i)\}_{i \in I}$$

be a chain of $\mathscr{A}$; let

$$B = \bigcup_{i \in I} B_i \quad \text{and} \quad G = \bigcup_{i \in I} G_i.$$

Then $(B, G) \in \mathscr{A}$.

Proof. By 1.40(i), $B \subseteq A$; thus, our result will be established if we can show that G well-orders B. First we verify that G is an order relation in B.

Reflexive. $x \in B \Rightarrow x \in B_i$ for some $i \in I \Rightarrow (x, x) \in G_i \subseteq G$; thus G is reflexive.

Antisymmetric. $(x, y) \in G$ and $(y, x) \in G \Rightarrow (x, y) \in G_i$ and $(y, x) \in G_j$ for some $i \in I$ and $j \in I$; but $\mathscr{C}$ is a chain of $\mathscr{A}$, so $G_i \subseteq G_j$ or $G_j \subseteq G_i$, say $G_i \subseteq G_j$. Thus $(x, y) \in G_j$ and $(y, x) \in G_j$; but G_j is an order relation, so $x = y$. This proves that G is antisymmetric.

Transitive. $(x, y) \in G$ and $(y, z) \in G \Rightarrow (x, y) \in G_i$ and $(y, z) \in G_j$ for some $i \in I$ and $j \in I$; but $\mathscr{C}$ is a chain, so $G_i \subseteq G_j$ or $G_j \subseteq G_i$, say $G_i \subseteq G_j$. Then $(x, y) \in G_j$ and $(y, z) \in G_j$, so $(x, z) \in G_j \subseteq G$. Thus G is transitive.

Now we must show that B is well-ordered by G. Suppose that $D \neq \varnothing$ and $D \subseteq B$; then $D \cap B_i \neq \varnothing$ for some $i \in I$. Now $D \cap B_i \subseteq B_i$, hence $D \cap B_i$ has a least element b in (B_i, G_i); that is, $\forall y \in D \cap B_i, (b, y) \in G_i$. We will proceed to show that b is the least element of D in (B, G); that is, $\forall x \in D, (b, x) \in G$.

Indeed, let $x \in D$: if $x \in B_i$, then $(b, x) \in G_i \subseteq G$. Now suppose $x \notin B_i$; in this case, $x \in B_j$ for some $j \in I$; $B_j \nsubseteq B_i$ because $x \in B_j$ and $x \notin B_i$, hence $(B_j, G_j) \nprec (B_i, G_i)$; it follows that $(B_i, G_i) \prec (B_j, G_j)$. Now we have $b \in B_i$, $x \in (B_j - B_i)$, and $(B_i, G_i) \prec (B_j, G_j)$; thus, by 5.19(c), $(b, x) \in G_j \subseteq G$. This proves that b is the least element of D in (B, G). ■

5.21 Lemma If $\mathscr{C}$, B, and G are defined as above, (B, G) is an upper bound of $\mathscr{C}$.

Proof. Let $(B_i, G_i) \in \mathscr{C}$; clearly $B_i \subseteq B$ and $G_i \subseteq G$. Now suppose that $x \in B_i$, $y \in B$, and $y \notin B_i$; certainly $y \in B_j$ for some $j \in I$. Now $B_j \nsubseteq B_i$ because $y \in B_j$ and $y \notin B_i$, so $(B_j, G_j) \nprec (B_i, G_i)$, hence $(B_i, G_i) \prec (B_j, G_j)$. Now $x \in B_i$ and $y \in (B_j - B_i)$, so by 5.19(c), $(x, y) \in G_j \subseteq G$. Thus $(B_i, G_i) \prec (B, G)$. ■

5.22 Theorem (*Well-ordering Theorem*). Any set A can be well ordered.

Proof. By lemmas 5.20 and 5.21, we can apply Zorn's Lemma to $\mathscr{A}$; thus $\mathscr{A}$ has a maximal element (B, G). We will show that $B = A$; hence A can be well-ordered. Otherwise, $\exists x \in (A - B)$; by defining x to be greater than each element of B, we get an extension G^* of G that well-orders $B \cup \{x\}$. (More explicitly, $G^* = G \cup \{(a, x) \mid a \in B\}$.) This is a contradiction, since (B, G) was assumed to be maximal. ∎

6 CONCLUSION

It is clear that the Axiom of Choice can be derived from the well-ordering theorem. Indeed, let A be any set; by the well-ordering theorem, A can be well ordered; if B is a nonempty subset of A, let $f(B)$ be the least element of B. Then f is a choice function on A.

We have now proven the following implications:

$$\text{Axiom of Choice} \Rightarrow \text{Hausdorff's Maximal Principle}$$
$$\Rightarrow \text{Zorn's Lemma} \Rightarrow \text{well-ordering theorem} \Rightarrow \text{Axiom of Choice.}$$

Thus we have established the complete equivalence of the above four propositions.

In the remainder of this book we will accept the Axiom of Choice as one of the axioms of set theory. Thus we will feel free to use the Axiom of Choice in all of our arguments, except if we make an explicit statement to the contrary.

6
The Natural Numbers

1 INTRODUCTION

Probably the most fundamental—and the most primitive—of all mathematical concepts is that of natural number. The natural, or "counting," numbers are the first mathematical abstraction which we learn as children; every human society—even the most backward and remote—possesses a system of some kind for counting objects.

We all have a clear intuitive understanding of what the natural numbers are: they are 0, 1, 2, 3, and so forth. But this intuitive perception—no matter how clear and immediate it may be—is not sufficient for the purposes of mathematics; in order to do mathematics with numbers, we must articulate this vague perception and transform it into a precise definition. The definition must, of course, faithfully reflect the intuitive notion from which it sprang.

It is our aim in this section to *define* the natural numbers; to be explicit, we will construct a set of objects to be called "natural numbers," and these "natural numbers" will be endowed with all of the properties which are associated with the natural numbers in our mind.

We should carefully note two important requirements of our definition. In the first place, we would like to present the natural numbers without introducing any new undefined notions. We were faced with a similar problem when we were about to introduce the notion of function; we solved it by defining a function to be a certain kind of class (specifically, a class of ordered pairs). Following this example, it is clear that we ought to define natural numbers to be classes (more specifically, *sets*) of some kind; in fact, for each n we will define "n" to be a set which (intuitively) has n elements. Secondly, each natural number must be uniquely defined; that is, for each n there must be just one, *unique* object which can be recognized as the "natural number n." Thus we must devise a means of fixing exactly one set, among all the sets with n elements, and calling this set "n."

The numbers "0," "1," "2," etc., which we are about to define will serve as standards in much the same way that the standard yard in Washington, D.C., serves as a norm for measuring length. It matters little whether the standard yard is made of platinum or stainless steel, or whether it is decorated with figures of dancing mermaids; its only use is as a standard of reference, so anything having the same length is, by definition, one yard long. Analogously, it does not matter too much which specific sets we define "0," "1," "2," etc., to be; they will be used as standards of reference, so a set will be said to "have n elements" if it is in one-to-one correspondence with the natural number "n."

We will proceed as follows: we define

$$0 = \varnothing.$$

In order to define "1," we must fix a set with exactly one element; thus

$$1 = \{0\}.$$

Continuing in this fashion, we define

$$2 = \{0, 1\},$$
$$3 = \{0, 1, 2\},$$
$$4 = \{0, 1, 2, 3\}, \quad \text{etc.}$$

The reader should note that $0 = \varnothing$, $1 = \{\varnothing\}$, $2 = \{\varnothing, \{\varnothing\}\}$, $3 = \{\varnothing, \{\varnothing\}, \{\varnothing, \{\varnothing\}\}\}$, etc. Our natural numbers are constructions beginning with the empty set.

The preceding definitions can be restated, a little more precisely, as follows. If A is a set, we define the *successor of A* to be the set A^+, given by

$$A^+ = A \cup \{A\}.$$

Thus, A^+ is obtained by adjoining to A exactly one new element, namely the element A. Now we define

$$0 = \varnothing,$$
$$1 = 0^+,$$
$$2 = 1^+,$$
$$3 = 2^+, \quad \text{etc.}$$

It is clear that these definitions coincide with those given in the preceding paragraph.

We have just outlined a method for producing sets which we call "natural numbers;" beginning with the empty set, we have given directions for constructing, successively, $0 = \varnothing$, $1 = 0^+$, $2 = 1^+$, and so forth. An important question now is the following: Is there such a thing as the *set of all the natural numbers* (or even the *class of all the natural numbers*)? That is, is there a set (or a class) which contains $\varnothing$, and which contains X^+ whenever it contains X? Certainly, our method does not enable us to construct it—we are merely given instructions for producing numbers 0, 1, 2, ..., *n up to any n*. Thus we cannot yet speak of the set (or class) of all the natural numbers.

A set A is called a *successor set* if it has the following properties:

i) $\varnothing \in A$.
ii) If $X \in A$, then $X^+ \in A$.

It is clear that any successor set necessarily includes all the natural numbers. Motivated by this observation, we introduce the following important axiom.

A9 (*Axiom of Infinity*). There exists a successor set.

As we have noted, every successor set includes all the natural numbers; thus it would make sense to define the "set of the natural numbers" to be the smallest successor set. Now it is easy to verify that any intersection of successor sets is a successor set; in particular, the intersection of all the successor sets is a successor set (it is obviously the smallest successor set). Thus, we are led naturally to the following definition.

6.1 Definition By the *set of the natural numbers* we mean the intersection of all the successor sets. The set of the natural numbers is designated by the symbol ω; every element of ω is called a *natural number.*

2 ELEMENTARY PROPERTIES OF THE NATURAL NUMBERS

In this section we will show that the natural numbers, as we have just defined them, satisfy the five conditions commonly known as the Peano axioms. This set of conditions—it is well known—is another way of defining and characterizing the natural numbers.

6.2 Theorem For each $n \in \omega$, $n^+ \neq 0$.

Proof. By definition, $n^+ = n \cup \{n\}$; thus $n \in n^+$ for each natural number n; but 0 is the empty set, hence 0 cannot be n^+ for any n. ■

6.3 Theorem (*Mathematical Induction*). Let X be a subset of ω; suppose X has the following properties:

i) $0 \in X$.
ii) If $n \in X$, then $n^+ \in X$.

Then $X = \omega$.

Proof. Conditions (i) and (ii) imply that X is a successor set. By 6.1, ω is a subset of every successor set; thus $\omega \subseteq X$. But $X \subseteq \omega$; so $X = \omega$. ■

6.4 Lemma Let m and n be natural numbers; if $m \in n^+$, then $m \in n$ or $m = n$.

Proof. By definition, $n^+ = n \cup \{n\}$; thus, if $m \in n^+$, then $m \in n$ or $m \in \{n\}$; but $\{n\}$ is a singleton, so $m \in \{n\}$ iff $m = n$. ■

6.5 Definition A set A is called *transitive* if, for each $x \in A$, $x \subseteq A$.

For example, the number 3 is a transitive set; indeed, its elements are 0, 1, 2, that is, $\varnothing$, $\{\varnothing\}$, $\{\varnothing, \{\varnothing\}\}$. It is clear that each of these elements is a subset of 3. The same is true of every natural number, as we shall prove next.

6.6 Lemma Every natural number is a transitive set.

Proof. Let X be the set of all the elements of ω which are transitive sets; we will prove, using mathematical induction (Theorem 6.3), that $X = \omega$; it will follow that every natural number is a transitive set.

i) $0 \in X$, for if 0 were not a transitive set, this would mean that $\exists y \in 0$ such that y is not a subset of 0; but this is absurd, since $0 = \varnothing$.

ii) Now suppose that $n \in X$; we will show that $n^+ \in X$; that is, assuming that n is a transitive set, we will show that n^+ is a transitive set. Let $m \in n^+$; by 6.4, $m \in n$ or $m = n$. If $m \in n$, then (because n is transitive) $m \subseteq n$; but $n \subseteq n^+$, so $m \subseteq n^+$. If $m = n$, then (because $n \subseteq n^+$) $m \subseteq n^+$; thus in either case, $m \subseteq n^+$, so $n^+ \in X$. It follows by 6.3 that $X = \omega$. ■

6.7 Theorem Let n and m be natural numbers. If $n^+ = m^+$, then $n = m$.

Proof. Suppose $n^+ = m^+$; now $n \in n^+$, hence $n \in m^+$; thus, by 6.4, $n \in m$ or $n = m$. By the very same argument, $m \in n$ or $m = n$. If $n = m$, the theorem is proved. Now suppose $n \neq m$; then $n \in m$ and $m \in n$. Thus, by 6.5 and 6.6, $n \subseteq m$ and $m \subseteq n$, hence $n = m$. ■

The Peano axioms for the natural numbers are:

P1 $0 \in \omega$.
P2 If $n \in \omega$, then $n^+ \in \omega$.
P3 For each $n \in \omega$, $n^+ \neq 0$.
P4 If X is a subset of ω such that
i) $0 \in X$, and
ii) if $n \in X$, then $n^+ \in X$,
then $X = \omega$.
P5 If $n, m \in \omega$ and $n^+ = m^+$, then $n = m$.

P1 and P2 follow immediately from our definition of ω. P3 is given by Theorem 6.2, P4 is given by Theorem 6.3, and P5 is given by Theorem 6.7. Thus our set ω satisfies the Peano axioms.

EXERCISES 6.2

1. Prove that A is a transitive set if and only if the following holds: If $B \in C$ and $C \in A$, then $B \in A$.
2. Prove that if A and B are transitive sets, then $A \cup B$ and $A \cap B$ are transitive sets.
3. Let A and B be sets. Prove that if $A = B$, then $A^+ = B^+$.
4. Use 6.3. to prove that for every natural number n, $n \notin n$.
5. Prove the following, where $m, n, p \in \omega$.

   a) $n \neq n^+$.
   b) If $m \in n$, then $n \notin m$.
   c) If $n \in m$ and $m \in p$, then $n \in p$.
   d) If $m \in n$, then $m^+ \subseteq n$.
6. a) Prove by induction: If $A \in n$ and $n \in \omega$, then $A \in \omega$. Conclude that ω is a transitive set.

   b) Prove that if $A^+ \in \omega$, then $A \in \omega$.
7. Prove that no natural number is a successor set.
8. Prove that no natural number is a subset of any of its elements.
9. Prove by induction: If $n \in \omega$, then either $n = 0$ or $n = m^+$ for some $m \in \omega$.
10. Let $n \in \omega$. Prove the following.

    a) $\cup n^+ = n$.
    b) $\cup \omega = \omega$ (see Remark 1.47).
11. Let A be a nonempty subset of ω. Prove that if $\cup A = A$ then $A = \omega$.

3 FINITE RECURSION

Induction is commonly used not only as a method of proof but also as a method of definition. For example, a familiar way of introducing exponents in arithmetic is by means of the "inductive definition"

I. $a^0 = 1$,
II. $a^{n+1} = a^n a$, $\forall n \in \omega$.

The pair of Conditions I and II is meant to be interpreted as a rule which specifies the meaning of a^n for each natural number n. Thus, by I, $a^0 = 1$; by I and II,

$$a^1 = a^{0+1} = a^0 a = 1a = a;$$

by II again,

$$a^2 = a^{1+1} = a^1 a = aa;$$

continuing in this fashion—using Condition II repeatedly—the numbers $a^0, a^1, a^2, \ldots, a^n$ are defined in succession up to any chosen n.

Inductive definitions such as the one we have just seen abound in mathematics. The situation, in almost every case, is the following: We have a set A, a function $f: A \to A$, and a fixed element $c \in A$. We define a function $\gamma: \omega \to A$ by means of the two Conditions

I. $\gamma(0) = c$,
II. $\gamma(n^+) = f(\gamma(n))$, $\forall n \in \omega$.

The reader should recognize that exactly this situation prevails in our preceding example. In that example, A is the set of the real numbers, c is 1, $\gamma(n)$ is denoted by a^n, and f is the function defined by $f(x) = xa$.

For another example, let $\mathbb{R}$ be the set of the real numbers and let $f: \mathbb{R} \to \mathbb{R}$ be the function defined by $f(x) = x^2$. We define a function $\gamma: \omega \to \mathbb{R}$ by

I. $\gamma(0) = 2$,
II. $\gamma(n + 1) = f(\gamma(n)) = [\gamma(n)]^2$.

The reader will recognize this as an inductive definition of the function $\gamma(n) = 2^{2^n}$.

In each of the foregoing examples, it is reasonable to believe that if γ exists, then Conditions I and II determine the values $\gamma(n)$ for every $n \in \omega$. However, Conditions I and II are insufficient in themselves to guarantee that a function such as γ exists. If we are to accept definition by induction as a legitimate way of constructing mathematical objects, then we must first establish the fact that γ—the function which we purport to be defining—actually exists and is uniquely determined. The purpose of the following theorem is to perform this important task.

6.8 Recursion Theorem Let A be a set, c a fixed element of A, and f a function from A to A. Then there exists a unique function $\gamma: \omega \to A$ such that

I. $\gamma(0) = c$, and
II. $\gamma(n^+) = f(\gamma(n))$, $\forall n \in \omega$.

Proof. First, we will establish the *existence* of γ. It should be carefully noted that γ is a set of ordered pairs which is a function and satisfies Conditions I and II. More specifically, γ is a subset of $\omega \times A$ with the following four properties:

1) $\forall n \in \omega, \exists x \in A \ni (n, x) \in \gamma$.
2) If $(n, x_1) \in \gamma$ and $(n, x_2) \in \gamma$, then $x_1 = x_2$.
3) $(0, c) \in \gamma$.
4) If $(n, x) \in \gamma$, then $(n^+, f(x)) \in \gamma$.

Properties (1) and (2) express the fact that γ is a function from ω to A, while properties (3) and (4) are clearly equivalent to I and II. We will now construct a graph γ with these four properties.

Let

$$\mathscr{A} = \{G \mid G \subseteq \omega \times A \text{ and } G \text{ satisfies (3) and (4)}\};$$

$\mathscr{A}$ is nonempty, because $\omega \times A \in \mathscr{A}$. It is easy to see that any intersection of elements of $\mathscr{A}$ is an element of $\mathscr{A}$; in particular,

$$\gamma = \bigcap_{G \in \mathscr{A}} G$$

is an element of $\mathscr{A}$. We proceed to show that γ is the function we require.

By construction, γ satisfies (3) and (4), so it remains only to show that (1) and (2) hold.

1) It will be shown by induction that dom $\gamma = \omega$, which clearly implies (1). By (3), $(0, c) \in \gamma$, so $0 \in \text{dom } \gamma$; now suppose $n \in \text{dom } \gamma$. Then $\exists x \in A \ni (n, x) \in \gamma$; by (4), then, $(n^+, f(x)) \in \gamma$, so $n^+ \in \text{dom } \gamma$. Thus, by Theorem 6.3, dom $\gamma = \omega$.

2) Let

$$N = \{n \in \omega \mid (n, x) \in \gamma \text{ for no more than one } x \in A\}.$$

It will be shown by induction that $N = \omega$. To prove that $0 \in N$, we first assume the contrary; that is, we assume that $(0, c) \in \gamma$ and $(0, d) \in \gamma$ where $c \neq d$. Let $\gamma^* = \gamma - \{(0, d)\}$; certainly γ^* satisfies (3); to show that γ^* satisfies (4), suppose that $(n, x) \in \gamma^*$. Then $(n, x) \in \gamma$, so $(n^+, f(x)) \in \gamma$; but $n^+ \neq 0$ (Theorem 6.2), so $(n^+, f(x)) \neq (0, d)$, and consequently $(n^+, f(x)) \in \gamma^*$. We conclude that γ^* satisfies (4), so $\gamma^* \in \mathscr{A}$; but γ is the intersection of all the elements of $\mathscr{A}$, so $\gamma \subseteq \gamma^*$. This is impossible, hence $0 \in N$. Next, we assume that $n \in N$ and prove that $n^+ \in N$. To do so, we first assume the contrary—that is, we suppose that $(n, x) \in \gamma$, $(n^+, f(x)) \in \gamma$, and $(n^+, u) \in \gamma$ where $u \neq f(x)$. Let $\gamma^\circ = \gamma - \{(n^+, u)\}$; y° satisfies (3) because $(n^+, u) \neq (0, c)$ (indeed, $n^+ \neq 0$ by Theorem 6.2). To show that γ° satisfies (4), suppose $(m, v) \in \gamma^\circ$; then $(m, v) \in \gamma$, so $(m^+, f(v)) \in \gamma$. Now we consider two cases, according as (a) $m^+ \neq n^+$ or (b) $m^+ = n^+$.

a) $m^+ \neq n^+$. Then $(m^+, f(v)) \neq (n^+, u)$, so $(m^+, f(v)) \in \gamma^\circ$.

b) $m^+ = n^+$. Then $m = n$ by 6.7, so $(m, v) = (n, v)$; but $n \in N$, so $(n, x) \in \gamma$ for no more than one $x \in A$; it follows that $v = x$, and so

$$(m^+, f(v)) = (n^+, f(x)) \in \gamma^\circ.$$

Thus, in either case (a) ot (b), $(m^+, f(v)) \in \gamma^\circ$; thus, γ° satisfies Condition (4), so $\gamma^\circ \in \mathscr{A}$. But γ is the intersection of all the elements of $\mathscr{A}$, so $\gamma \subseteq \gamma^\circ$; this is impossible, so we conclude that $n^+ \in N$. Thus $N = \omega$.

Finally, we will prove that γ is *unique*. Let γ and γ' be functions, from ω to A which satisfy I and II. We will prove by induction that $\gamma = \gamma'$. Let

$$M = \{n \in \omega \mid \gamma(n) = \gamma'(n)\}.$$

Now $\gamma(0) = c = \gamma'(0)$, so $0 \in M$; next, suppose that $n \in M$. Then

$$\gamma(n^+) = f(\gamma(n)) = f(\gamma'(n)) = \gamma'(n^+),$$

hence $n^+ \in M$. ■

6.9 Corollary Let f, c, and γ be as in Theorem 6.8. If f is injective and $c \notin \operatorname{ran} f$, then γ is injective.

Proof. We wish to show that if $\gamma(m) = \gamma(n)$, then $m = n$; the proof is by induction on m.

i) $m = 0$. If $n = 0$ we are done; if $n \neq 0$, then $n = k^+$ for some $k \in \omega$, so

$$c = \gamma(0) = \gamma(m) = \gamma(n) = \gamma(k^+) = f(\gamma(k)),$$

which is impossible because c is not in the range of f. Thus $n = 0 = m$.

ii) Suppose the corollary is true for m; let $\gamma(m^+) = \gamma(n)$. If $n = 0$ then we have $\gamma(0) = \gamma(m^+)$, which, as we have just shown, is impossible; thus $n \neq 0$, so $n = k^+$ for some $k \in \omega$. Thus $\gamma(m^+) = \gamma(k^+)$, that is, $f(\gamma(m)) = f(\gamma(k))$; but f is injective, so $\gamma(m) = \gamma(k)$. By the hypothesis of induction, it follows that $m = k$; hence $m^+ = k^+ = n$. ■

EXERCISES 6.3

1. Let $a \in \mathbb{R}$, where $\mathbb{R}$ is the set of the real numbers; define a^n by the following two conditions.

$$a^0 = 1,$$
$$a^{n+1} = a^n a, \quad \forall n \in \omega.$$

Prove that for each $n \in \omega$, a^n is a uniquely defined real number. (In other words, prove that $\gamma(n) = a^n$ is a uniquely determined function $\omega \to \mathbb{R}$; use Theorem 6.8.)

2. Let A be a set and let $f: A \to A$ be a function. Define f^n by

$$f^0 = I_A \qquad \text{(the identity function on } A\text{)},$$
$$f^{n+1} = f^n \circ f, \quad \forall n \in \omega.$$

Prove that for each $n \in \omega$, f^n is a uniquely determined element of A^A.

3. Let A be a set and let $f: A \to B$ be an injective function, where $B \subset A$. Prove that A has a subset which is in one-to-one correspondence with ω. [*Hint:* Use 6.9 to prove that there is an injective $\gamma: \omega \to D$ where $D \subseteq A$.]

4. If A is a partially ordered set, by a *strictly increasing sequence* in A we mean a function $\gamma: \omega \to A$ such that $\gamma(0) < \gamma(1) < \gamma(2) < \cdots$ Let A be a partially ordered set which has no maximal elements; prove that there is a strictly increasing sequence in A.

4 ARITHMETIC OF NATURAL NUMBERS

One of the most important applications of the recursion theorem is its use in defining addition and multiplication of natural numbers.

If m is a natural number, the recursion theorem guarantees the existence of a unique function $\gamma_m: \omega \to \omega$ defined by the two Conditions

I. $\gamma_m(0) = m$,
II. $\gamma_m(n^+) = [\gamma_m(n)]^+, \forall n \in \omega$.

Addition of natural numbers is now defined as follows:

$$m + n = \gamma_m(n)$$

for all $m, n \in \omega$. Conditions I and II immediately above can be rewritten thus:

6.10
$$m + 0 = m,$$
$$m + n^+ = (m + n)^+.$$

We proceed to derive a few simple properties of addition.

6.11 Lemma $n^+ = 1 + n$, where 1 is defined to be 0^+.

Proof. This can be proven by induction on n. If $n = 0$, then we have

$$0^+ = 1 = 1 + 0$$

(this last equality follows from 6.10), hence the lemma holds for $n = 0$. Now, assuming the lemma is true for n, let us show that it holds for n^+:

$$\begin{aligned} 1 + n^+ &= (1 + n)^+ && \text{by 6.10} \\ &= (n^+)^+ && \text{by the hypothesis of induction. } \blacksquare \end{aligned}$$

6.12 Lemma $0 + n = n$.

Proof. Let $X = \{n \in \omega \mid 0 + n = n\}$; it will be shown by induction that $X = \omega$. Indeed, $0 + 0 = 0$ by 6.10, hence $0 \in X$. Now suppose that $n \in X$, that is, $0 + n = n$. Then

$$\begin{aligned} 0 + n^+ &= (0 + n)^+ && \text{by 6.10} \\ &= n^+ && \text{by the hypothesis of induction.} \end{aligned}$$

It follows by Theorem 6.3 that $X = \omega$. $\blacksquare$

6.13 Theorem $(m + n) + k = m + (n + k)$.

Proof. The proof is by induction. For arbitrary elements $m, n \in \omega$, let

$$L_{mn} = \{k \in \omega \mid (m + n) + k = m + (n + k)\};$$

it will be shown that $L_{mn} = \omega$. First,

$$(m + n) + 0 = m + n = m + (n + 0);$$

hence $0 \in L_{mn}$. Now suppose $k \in L_{mn}$, that is,

$$(m + n) + k = m + (n + k).$$

Then

$$\begin{aligned} (m + n) + k^+ &= ((m + n) + k)^+ && \text{by 6.10} \\ &= (m + (n + k))^+ && \text{because } k \in L_{mn} \\ &= m + (n + k)^+ && \text{by 6.10} \\ &= m + (n + k^+) && \text{by 6.10,} \end{aligned}$$

so $k^+ \in L_{mn}$. ■

6.14 Theorem $m + n = n + m$.

Proof. For an arbitrary natural number m, let

$$L_m = \{n \in \omega \mid m + n = n + m\}.$$

It will be proven by induction that $L_m = \omega$. Now $m + 0 = m = 0 + m$ by 6.10 and 6.12, hence $0 \in L_m$. Next suppose $n \in L_m$, that is, $m + n = n + m$. Then

$$\begin{aligned} m + n^+ &= (m + n)^+ && \text{by 6.10} \\ &= (n + m)^+ && \text{by the hypothesis of induction} \\ &= 1 + (n + m) && \text{by 6.11} \\ &= (1 + n) + m && \text{by 6.13} \\ &= n^+ + m && \text{by 6.11.} \end{aligned}$$ ■

If m is a natural number, the recursion theorem guarantees the existence of a unique function $\beta_m : \omega \to \omega$ defined by the two Conditions

I. $\beta_m(0) = 0$,
II. $\beta_m(n^+) = \beta_m(n) + m, \ \forall n \in \omega$.

Multiplication of natural numbers is now defined as follows:

$$mn = \beta_m(n)$$

for all $m, n \in \omega$. Conditions I and II immediately above can be rewritten thus:

6.15 $$m0 = 0,$$
$$mn^+ = mn + m.$$

The following are a few simple properties of multiplication.

6.16 Lemma . $0n = 0$.

Proof. Let $N = \{n \in \omega \mid 0n = 0\}$; it will be shown by induction that $N = \omega$. Indeed, $00 = 0$ by 6.15, hence $0 \in N$. Now suppose that $n \in N$, that is, $0n = 0$. Then by 6.15, 6.10, and the hypothesis of induction,

$$0n^+ = 0n + 0 = 0n = 0;$$

hence $n^+ \in N$. It follows by induction that $N = \omega$. ■

6.17 Lemma $1n = n$.
The proof (by induction) is left as an exercise for the reader.

6.18 Theorem (*Distributive Law*).
i) $m(n + k) = mn + mk$.
ii) $(n + k)m = nm + km$.

Proof
i) If m and n are natural numbers, let

$$L_{mn} = \{k \in \omega \mid m(n + k) = mn + mk\}.$$

It will be shown by induction that $L_{mn} = \omega$. Now

$$m(n + 0) = mn = mn + 0 = mn + m0,$$

hence $0 \in L_{mn}$. Next, suppose that $k \in L_{mn}$, that is, $m(n + k) = mn + mk$; then, by 6.10, 6.13 and 6.15,

$$m(n + k^+) = m(n + k)^+ = m(n + k) + m$$
$$= (mn + mk) + m = mn + (mk + m) = mn + mk^+,$$

hence $k^+ \in L_{mn}$.

ii) The proof is left as an exercise for the reader. ■

6.19 Theorem (*Associative Law for Multiplication*). $(mn)k = m(nk)$.

Proof. Let $L_{mn} = \{k \in \omega \mid (mn)k = m(nk)\}$; it will be proved by induction that $L_{mn} = \omega$. First, by 6.15,

$$(mn)0 = 0 = m0 = m(n0),$$

hence $0 \in L_{mn}$. Next, assume that $k \in L_{mn}$, that is, $(mn)k = m(nk)$. Then, by 6.15 and 6.18,

$$\begin{aligned}(mn)k^+ &= (mn)k + mn = m(nk) + mn \\ &= m(nk + n) = m(nk^+);\end{aligned}$$

hence $k^+ \in L_{mn}$. ■

6.20 Theorem (*Commutative Law for Multiplication*). $mn = nm$.

The proof is left as an exercise for the reader.

One of the most important aspects of the natural numbers is their ordering. Before proceeding with a formal definition of the order relation in ω, the reader should review the definition of ω. Specifically, it should be noted that for each n, the natural number n is the set of all the natural numbers preceding n:

$$n = \{0, 1, \ldots, n - 1\}.$$

It is clear, now, how we are to define order in ω: n is to precede m if and only if n is an element of m. Motivated by this observation, we make the following formal definition.

6.21 Definition A relation $\leqslant$ is defined in ω as follows:

$$m \leqslant n \text{ if and only if } m \in n \text{ or } m = n.$$

First, it is required to prove that this is indeed an order relation in ω.

6.22 Theorem Let $m \leqslant n$ denote the fact that $m \in n$ or $m = n$. Then the relation $\leqslant$ is an order relation in ω.

Proof

i) For each $m \in \omega$, $m = m$, hence $m \leqslant m$ (reflexive law).
ii) Suppose $m \leqslant n$ and $n \leqslant m$; this means that either $m = n$, or $m \in n$ and $n \in m$. In the latter case, $m \subseteq n$ and $n \subseteq m$ by 6.6, hence again $m = n$ (antisymmetric law).
iii) Suppose $m \leqslant n$ and $n \leqslant p$; we have four possible cases.
   1. $m \in n$ and $n \in p$: thus $m \in n$ and $n \subseteq p$, so $m \in p$.
   2. $m \in n$ and $n = p$: thus $m \in p$.
   3. $m = n$ and $n \in p$: thus $m \in p$.
   4. $m = n$ and $n = p$: thus $m = p$.

In each case, $m \leqslant p$ (transitive law). ■

We will show next that ω is well ordered; this will require the following two lemmas.

6.23 Lemma If m is a natural number, $0 \leqslant m$.

Proof. Let $L = \{m \in \omega \mid 0 \leqslant m\}$; by the reflexive law 6.22(i), $0 \leqslant 0$, so $0 \in L$. Now suppose $m \in L$, that is, $0 \leqslant m$. From $m \in m^+$ it follows that $m \leqslant m^+$; thus by the transitive law 6.22(iii), $0 \leqslant m^+$, so $m^+ \in L$. So by 6.3, $L = \omega$. ■

6.24 Lemma If $n < m$ then $n^+ \leqslant m$.

Proof. If n is a natural number, let

$$L_n = \{m \in \omega \mid n < m \Rightarrow n^+ \leqslant m\};$$

We will use induction (6.3) to prove that $L_n = \omega$. Note that $m \notin L_n$ iff $n < m$ and $n^+ \nleqslant m$, that is, $n \in m$ and $n^+ \nleqslant m$. In particular, $0 \notin L_n$ iff $n \in 0$ and $n^+ \nleqslant 0$, which is impossible (specifically, $n \in 0$ is impossible); thus $0 \in L_n$. Now assume that $m \in L_n$, that is, $n < m \Rightarrow n^+ \leqslant m$, and let us show that $m^+ \in L_n$, that is,

$$n < m^+ \Rightarrow n^+ \leqslant m^+.$$

If $n < m^+$, that is, $n \in m^+$, then by 6.4, $n \in m$ or $n = m$. If $n = m$, then $n^+ = m^+$, and we are done. If $n \in m$, that is, $n < m$, then by the hypothesis of induction, $n^+ \leqslant m < m^+$, so we are done again. ■

6.25 Theorem ω is well ordered.

Proof. Suppose, on the contrary, that A is a nonempty subset of ω without a least element. Let

$$L = \{n \in \omega \mid n \leqslant m \text{ for every } m \in A\}.$$

By 6.23, $0 \in L$. Now suppose that $n \in L$, that is, $n \leqslant m$ for every $m \in A$. If $n = p$ for some $p \in A$, then p is the least element of A, contrary to our hypothesis; thus $n < m$ for every $m \in A$. It follows by 6.24 that $n^+ \leqslant m$ for every $m \in A$, so $n^+ \in L$. Thus by 6.3, $L = \omega$. But $L \cap A = \varnothing$ because A has no least element. Thus $A = \varnothing$. ■

EXERCISES 6.4

1. Prove Theorem 6.18(ii).
2. Prove Theorem 6.20.
3. Prove each of the following.
   a) $m = n \Rightarrow m + k = n + k$ (see Exercise 3, Exercise Set 6.2).
   b) $m = n \Rightarrow mk = nk$.
4. Prove each of the following.
   a) $m < 1 \Rightarrow m = 0$.
   b) There is no natural number k such that $m < k < m^+$.

5. Prove each of the following.
   a) $n < k \Rightarrow m + n < m + k$.
   b) $m + n = m + k \Rightarrow n = k$.
6. Prove each of the following.
   a) If $m < n$ and $k \neq 0$, then $mk < nk$.
   b) If $mk = nk$ and $k \neq 0$, then $m = n$.
7. Prove that if $m \leqslant n$, then there exists a unique $p \in \omega$ such that $m + p = n$.
8. Prove each of the following.
   a) $m + k < n + k \Rightarrow m < n$.
   b) $mk < nk \Rightarrow m < n$.
9. Give an inductive definition of exponentiation of natural numbers; that is, define m^n in a manner similar to 6.10 and 6.15, justifying your definition in terms of the recursion theorem. Then prove each of the following.
   a) $m^{n+k} = m^n m^k$, b) $(mn)^k = m^k n^k$, c) $(m^n)^k = m^{nk}$.

5 CONCLUDING REMARKS

By Axiom A9, there exists a successor *set* X; by 6.1, $\omega \subseteq X$, hence by Axiom A3, *ω is a set*. It follows, by Axiom A3 again, that *every natural number is a set.*

In the next chapter we will define a class to be *finite* if it is in one-to-one correspondence with a natural number; it follows by 2.36 that *every finite class is a set.* In view of this remark, we will henceforth speak of finite *sets* rather than finite classes.

We have just seen that ω is a set; thus, by 1.53, $\omega \times \omega$ is a set. Now, if we identify each fraction n/m (where n/m is assumed to be in "lowest terms") with the ordered pair (n, m), then the class of all the positive rational numbers is a subclass of $\omega \times \omega$, hence by A3, it is a set. Analogously, the class of all the negative rational numbers is a set, hence by A5, the class $\mathbb{Q}$ of all the rational numbers is a set. It is well known that every real number can be regarded as a sequence (called a *Cauchy sequence*) of rational numbers; hence, roughly speaking,* as an element of $\mathscr{P}(\mathbb{Q})$. In other words, the class $\mathbb{R}$ of the real numbers is a subclass of $\mathscr{P}(\mathbb{Q})$; hence, by A6 and A3, it is a set. In similar fashion, the class $\mathbb{C}$ of the complex numbers is a set.

We have seen in preceding sections that the union of any set of sets is a set; if A is a set, then $\mathscr{P}(A)$ is a set; if $\{A_i\}_{i \in I}$ is a family of sets, where I is a set, then $\prod_{i \in I} A_i$ is a set; and if A and B are sets, then A^B is a set. Thus it is clear that every object we can produce by the classical construction processes is a set.

All the infinite (that is, not finite) classes which occur in traditional mathematics are sets; thus, in the next chapter, we will confine our attention to finite and infinite *sets*.

---

* To be more precise, every Cauchy sequence is an element of $\mathbb{Q}^\omega$.

7
Finite and Infinite Sets

1 INTRODUCTION

One of the most fundamental distinctions in mathematics is that between finite and infinite sets. The distinction is so intuitively compelling that, even in the absence of a precise definition, there cannot be any doubt as to whether a given set is finite or infinite. In simple terms, a finite set is one which "has n elements," where n is a natural number, and an infinite set is one which is not finite.

Although the dichotomy between finite and infinite has always fascinated mathematicians—it has been the source of the most celebrated riddles, paradoxes, and classical errors of mathematics—a sound theory of infinite sets did not appear until very recent times. It had to await the arrival of the rigorous concepts of set, mapping, and one-to-one correspondence. Once the use of these new tools became familiar to mathematicians, toward the end of the nineteenth century, the modern theory of infinite sets developed rapidly; it was largely the work of Georg Cantor and his successors.

Using familiar concepts, and arguments which are remarkable for their simplicity, Cantor was able to draw conclusions which surprised mathematicians and laymen alike. Cantor's ideas are well known today; they have been popularized in innumerable expository books and articles, and have entered the lore of modern mathematics. We proceed, in the remainder of this section, to give the bare outlines of Cantor's theory. In this discussion the words "finite" and "infinite" will be used informally; as we remarked earlier, a finite set can be described as one which "has n elements" (n is a natural number), and an infinite set is, simply, one which is not finite.

Two finite sets A and B have the "same number of elements" if and only if they are in one-to-one correspondence. Even though we cannot speak of two infinite sets as having the "same number of elements," we have the feeling, nonetheless, that if A and B are infinite sets and there is a one-to-one correspondence between them, then, in a certain sense, they are of the "same size." This intuitive notion is formalized by defining two sets A and B to be *equipotent*, or to have the *same power*, if there is a one-to-one correspondence from A to B. We say that A is of a *lesser power* than B if there exists a one-to-one correspondence between A and a proper subset of B, but none between A and

B. Here, again, we have taken our cue from the finite case: for if A and B are finite sets and A has fewer elements than B, then certainly there exists a one-to-one correspondence between A and a part of B, but none between A and all of B. If A and B have the same power, we write $A \approx B$; if the power of A is less than that of B, we write $A \prec B$; finally, if the power of A is less than or equal to that of B (that is, A is in one-to-one correspondence with a subset of B), then we write $A \preccurlyeq B$.

It is a curious fact, first proven by Cantor, that many sets which appear to be smaller—or larger—than ω actually have the same power as ω. For example, if E is the set of the even natural numbers, it is easy to see that the function $f(n) = 2n$ is a bijective function from ω to E. Thus, although E is a proper subset of ω (in fact, E appears to have only "half as many" elements as ω), actually E is equipotent with ω. A more surprising example involves ω and the set $\mathbb{Q}$ of the rational numbers; our intuition suggests, in the most compelling way, that $\mathbb{Q}$ is a "larger" set than ω; for $\mathbb{Q}$ not only includes ω, but is, in an obvious manner, "infinitely dense" with respect to ω. Yet it can easily be shown that ω and $\mathbb{Q}$ are in one-to-one correspondence; the proof consists in "enumerating" the rational numbers—that is, making a list $r_1, r_2, r_3, \ldots$ of rational numbers which includes them all; the correspondence $i \leftrightarrow r_i$ is then a one-to-one correspondence between ω and $\mathbb{Q}$. We proceed as follows.

First, we group all the positive rational numbers into classes $A_1, A_2, \ldots,$ where A_i contains all the fractions n/m such that $n + m = i$. Within each class A_i, we order the numbers n/m in increasing order of the numerator n. Thus the first few fractions in this ordering would be as follows:

$$\tfrac{1}{1}, \tfrac{1}{2}, \tfrac{2}{1}, \tfrac{1}{3}, \tfrac{2}{2}, \tfrac{3}{1}, \tfrac{1}{4}, \tfrac{2}{3}, \tfrac{3}{2}, \tfrac{4}{1}, \tfrac{1}{5}, \ldots.$$

Now we delete all fractions which are not in "lowest terms;" this leaves

$$1, \tfrac{1}{2}, 2, \tfrac{1}{3}, 3, \tfrac{1}{4}, \tfrac{2}{3}, \tfrac{3}{2}, \ldots.$$

This is clearly an enumeration $r_1, r_2, \ldots$ of the positive rational numbers. If $t_1, t_2, \ldots$ is a similar enumeration of the negative rational numbers, then $0, r_1, t_1, r_2, t_2, \ldots$ is an enumeration of all the rational numbers.

A set which is in one-to-one correspondence with ω is said to be *denumerable*. Faced with the unexpected discovery that $\mathbb{Q}$ is denumerable, we are naturally led to wonder whether every infinite set is denumerable. This question was answered by Cantor—in the negative: the set of all the real numbers, for example, is not denumerable. To prove this, we use the so-called *diagonal method*.

First, we note that the function $y = \tan(\pi x - \pi/2)$ is a one-to-one correspondence between the set of all the real numbers and the open interval $(0, 1)$; hence it will be sufficient for our purposes to prove that the set of all the real numbers r such that $0 < r < 1$ cannot be enumerated. We argue by contradiction. Suppose that $r_1, r_2, \ldots$ is an enumeration of all the real numbers between 0 and 1. Let each real number be expressed as a nonterminating decimal; thus $r_i = \cdot r_{i1} r_{i2} r_{i3} \ldots$, where each r_{ij} is a digit 0, 1, ..., 9.

$$\begin{aligned} r_1 &= \cdot r_{11} r_{12} r_{13} r_{14} \cdots \\ r_2 &= \cdot r_{21} r_{22} r_{23} r_{24} \cdots \\ r_3 &= \cdot r_{31} r_{32} r_{33} r_{34} \cdots \\ r_4 &= \cdot r_{41} r_{42} r_{43} r_{44} \cdots \\ &\cdot \\ &\cdot \\ &\cdot \end{aligned}$$

Now we define s to be a number $s = \cdot s_1 s_2 s_3 \ldots$, where $s_1 \neq r_{11}$, $s_2 \neq r_{22}$, $s_3 \neq r_{33}$, $s_4 \neq r_{44}$, and so forth (for example, we might dictate that $s_i = 1$ if $r_{ii} \neq 1$ and $s_i = 2$ if $r_{ii} = 1$). Now, since $0 < s < 1$, it follows that s is one of the numbers in the enumeration, say $s = r_k$. But this is impossible, because the kth digit of s is s_k and the kth digit of r_k is r_{kk}, and $s_k \neq r_{kk}$. Because of this contradiction, it is clear that there is no way of enumerating the real numbers. Yet ω has the same power as a subset of the real numbers—namely those real numbers which happen to be positive integers. Thus $\omega \prec \mathbb{R}$.

We have just revealed one of the most significant facts in Cantor's theory of the infinite: while ω and $\mathbb{R}$ are both infinite sets, one of them is strictly larger than the other; in other words infinite sets, like finite sets, come in different "sizes." Our next step, naturally, is to find a set which is strictly larger than $\mathbb{R}$. In order to settle this question, however, we introduce a result of far greater generality, which will provide us at once with a strictly increasing sequence $K_1 \prec K_2 \prec K_3 \prec \ldots$ of infinite sets. Our result is simply this: if A is any set, then the power set of A, $\mathscr{P}(A)$, is strictly larger than A. To prove this, we use a variant of the diagonal method which served us in the preceding paragraph.

We begin by assuming that there *is* a one-to-one correspondence $\phi: A \to \mathscr{P}(A)$. We define a set B as follows:

$$B = \{x \in A \mid x \notin \phi(x)\};$$

B is a subset of A, so $B = \phi(y)$ for some $y \in A$. Now if $y \in \phi(y)$, then $y \notin B$, that is, $y \notin \phi(y)$; yet if $y \notin \phi(y)$, then $y \in B$, that is, $y \in \phi(y)$. Quite obviously this is impossible; hence there exists no one-to-one correspondence between A and $\mathscr{P}(A)$. However, A has the same power as a subset of $\mathscr{P}(A)$, namely the set of all the singletons $\{x\}$. We conclude that $A \prec \mathscr{P}(A)$.

The argument of the preceding paragraph is a proof for the following theorem.

7.1 Theorem If A is a set, there exists no surjective function $A \to \mathscr{P}(A)$.

7.2 Corollary No subset of A can be equipotent with $\mathscr{P}(A)$.

7.3 Corollary A cannot be equipotent with any set containing $\mathscr{P}(A)$.

It follows from 7.2 and 7.3 that

7.4 if $B \subseteq A$, then $B \prec \mathscr{P}(A)$;
if $\mathscr{P}(A) \subseteq D$, then $A \prec D$.

We have proved earlier (Theorem 2.35) that if A is a set, $\mathscr{P}(A)$ and 2^A are in one-to-one correspondence. Thus all of the above statements hold true when we replace $\mathscr{P}(A)$ by 2^A.

If we let $\omega = K_1$, $\mathscr{P}(K_1) = K_2$, $\mathscr{P}(K_2) = K_3$, and so forth, then we have the strictly increasing sequence of infinite sets

$$K_1 \prec K_2 \prec K_3 \prec \cdots$$

Now consider $L_1 = \bigcup_{i \in \omega} K_i$; for each i, $K_{i+1} \subseteq L_1$, that is, $\mathscr{P}(K_i) \subseteq L_1$. It follows by 7.4 that for each i, $K_i \prec L_1$. Now we let $L_2 = \mathscr{P}(L_1)$, $L_3 = \mathscr{P}(L_2)$, and so forth; hence we have the strictly increasing sequence of infinite sets

$$K_1 \prec K_2 \prec K_3 \prec \cdots \prec L_1 \prec L_2 \prec L_3 \prec \cdots \prec M_1 \prec M_2 \prec M_3 \prec \cdots$$

Thus, speaking informally, there are many more "sizes" of infinite sets than there are different "sizes" of finite sets.

It is worth noting that the set $\mathbb{R}$ of the real numbers is equipotent with 2^ω. Indeed, we have already noted that $\mathbb{R}$ is equipotent with the open interval $(0, 1)$ of $\mathbb{R}$; now each element r in the interval $(0, 1)$ can be written in binary notation

$$r = .r_1 r_2 r_3 \ldots,$$

where each r_i is either 0 or 1. This expression for r can be identified with the function $\phi_r\colon \omega \to \{0, 1\}$ given by $\phi_r(i) = r_i$, $\forall i \in \omega$. It is easy to see that the correspondence $r \leftrightarrow \phi_r$ is a one-to-one correspondence between the interval $(0, 1)$ and 2^ω.

We have seen that the set $\mathbb{Q}$ of the rational numbers is equipotent with ω; it is easy to show that the set $\mathbb{C}$ of the complex numbers is equipotent with $\mathbb{R}$; thus, by the preceding paragraph, all of classical mathematics deals with only two sizes of infinite sets, namely, sets equipotent with ω and sets equipotent with 2^ω; the power of 2^ω is often called the *power of the continuum.* Now an interesting question which arises is the following: is there a power between that of ω and that of 2^ω? That is, does there exist any set A such that $\omega \prec A \prec 2^\omega$? Since no such set occurs anywhere in classical mathematics, and there appears to be no way of constructing one, it was conjectured by Cantor and his contemporaries that the answer to that question must be "no;"

this conjecture is known as the *continuum hypothesis.* A closely related conjecture is the *generalized continuum hypothesis,* which proposes that for every set B, there is no set A such that $B \prec A \prec 2^B$. These hypotheses have never been either proven or disproven; we shall have more to say about them in Chapter 9.

The aim of this chapter is to exploit the various ideas which have been motivated in our introduction. We will define rigorously the concepts of finite and infinite set, "power," and cardinality, and give the classical results of Cantor's theory.

2 EQUIPOTENCE OF SETS

In the preceding section, we have defined the symbols $\approx$, $\prec$, and $\preccurlyeq$ as follows:

$A \approx B$ iff A is in one-to-one correspondence with B.

$A \preccurlyeq B$ iff A is in one-to-one correspondence with a subset of B.

$A \prec B$ iff A is in one-to-one correspondence with a subset of B and A is not in one-to-one correspondence with B.

It is immediate from the second of these statements that

7.5 $A \preccurlyeq B$ iff there exists an injective function $A \to B$.

Furthermore, we have the following.

7.6 Lemma There exists an injective function $f: A \to B$ if and only if there exists a surjective function $g: B \to A$.

Proof

i) Suppose $f: A \to B$ is injective; by 2.25, there exists a function $g: B \to A$ such that $g \circ f = I_A$; thus, by 5.4, g is surjective.

ii) Suppose $g: B \to A$ is surjective; by 5.4, there exists a function $f: A \to B$ such that $g \circ f = I_A$; thus, by 2.25, f is injective. ■

By 7.5 and 7.6 we have

7.7 $A \preccurlyeq B$ iff there exists a surjective function $B \to A$.

7.8 Theorem Let A, B, C, and D be sets where $A \cap C = \varnothing$ and $B \cap D = \varnothing$. If $f: A \to B$ and $g: C \to D$ are bijective functions, then $f \cup g$ is a bijective function $A \cup C \to B \cup D$.

Proof. If $f: A \to B$ and $g: C \to D$ are functions, then clearly $f: A \to B \cup D$

and $g: C \to B \cup D$ are functions, hence by 2.16,

$$(f \cup g): A \cup C \to B \cup D$$

is a function. Now $f: A \to B$ and $g: C \to D$ are bijective, hence by 2.21,

$$f^{-1}: B \to A \qquad \text{and} \qquad g^{-1}: D \to C$$

are functions; thus, as above,

$$(f^{-1} \cup g^{-1}): B \cup D \to A \cup C$$

is a function. But clearly $f^{-1} \cup g^{-1} = (f \cup g)^{-1}$, hence

$$(f \cup g)^{-1}: B \cup D \to A \cup C$$

is a function; thus, by 2.22,

$$(f \cup g): A \cup C \to B \cup D$$

is bijective. ■

7.9 Corollary Suppose $A \cap C = \varnothing$ and $B \cap D = \varnothing$; if $A \approx B$ and $C \approx D$, then $A \cup C \approx B \cup D$.

7.10 Theorem If $A \approx B$ and $C \approx D$, then $A \times C \approx B \times D$.

Proof. Let $f: A \to B$ and $g: C \to D$ be bijective functions, and let us define $h: A \times C \to B \times D$ as follows:

$$h(x, y) = \big(f(x), g(y)\big), \quad \forall (x, y) \in A \times C.$$

It can easily be shown that $h: A \times C \to B \times D$ is bijective; the details are left to the reader. ■

7.11 Theorem If $A \approx B$ and $C \approx D$, then $A^C \approx B^D$.

Proof. Let $f: A \to B$ and $g: D \to C$ be bijective functions, and let us define $h: A^C \to B^D$ in the following way. For each $\alpha \in A^C$, that is, for each function $\alpha: C \to A$, let $h(\alpha) = f \circ \alpha \circ g$; clearly $f \circ \alpha \circ g$ is a function $D \to B$, that is, $f \circ \alpha \circ g \in B^D$. It can be shown routinely that $h: A^C \to B^D$ is a bijective function; the details are left to the reader. ■

7.12 Corollary If $A \approx B$, then $\mathscr{P}(A) \approx \mathscr{P}(B)$.

This follows immediately from 7.11 and 2.35.

EXERCISES 7.2

1. Complete the proof of Theorem 7.10.
2. Complete the proof of Theorem 7.11.
3. Prove that if $(A - B) \approx (B - A)$, then $A \approx B$.
4. Suppose $A \approx B$, $a \in A$, and $b \in B$. Prove that $(A - \{a\}) \approx (B - \{b\})$.
5. Suppose that $A \approx B$, $C \approx D$, $C \subset A$ and $D \subset B$. Prove that $(A - C) \approx (B - D)$.
6. Let $\{B_i\}_{i \in I}$ and $\{C_i\}_{i \in I}$ each be a family of mutually disjoint sets. If $B_i \approx C_i$ for each $i \in I$, prove that

$$\bigcup_{i \in I} B_i \approx \bigcup_{i \in I} C_i.$$

7. Let $\{B_i\}_{i \in I}$ and $\{C_i\}_{i \in I}$ be families of sets. If $B_i \approx C_i$ for each $i \in I$, prove that

$$\prod_{i \in I} B_i \approx \prod_{i \in I} C_i.$$

3 PROPERTIES OF INFINITE SETS

A set A is said to be *finite* if A is in one-to-one correspondence with a natural number n; otherwise, A is said to be *infinite*. Several other definitions of "finite" and "infinite" are to be found in the mathematical literature; foremost among them are the following:

i) A is infinite if and only if A has a denumerable subset.
ii) A is infinite if and only if A is equipotent with a proper subset of itself.

In each of the above two cases, a set is called "finite" if it is not infinite.

It will be shown next that (i) and (ii) are each equivalent to our definition of "infinite," given above.

7.13 Lemma If A is a denumerable set and $x \in A$, then $A - \{x\}$ is a denumerable set.

Proof. If A is denumerable, then there exists a bijective function $f: \omega \to A$. Corresponding to x, there is an $n \in \omega$ such that $f(n) = x$; define $g: \omega \to A$ as follows:

$$g(m) = \begin{cases} f(m) & \text{if } m < n, \\ f(m+1) & \text{if } m \geqslant n. \end{cases}$$

It is easy to see that g is a bijective function from ω to $A - \{x\}$; the details are left to the reader. ■

7.14 Theorem A is an infinite set if and only if A has a denumerable subset.

Proof

i) Well-order A; by Theorem 4.62, exactly one of the following cases holds: (α) ω is isomorphic with A; (β) ω is isomorphic with an initial segment of A; (γ) A is isomorphic with an initial segment of ω. If A does not have a denumerable subset, then (α) and (β) cannot hold, hence (γ) holds; therefore A is equipotent with an initial segment $n = S_n$ of ω, so A is finite.* We have just proved that if A does *not* have a denumerable subset, then A is finite.

ii) To prove the converse, we will first use induction to show that a natural number n cannot have a denumerable subset. This assertion is clearly true for $n = 0$; let it be true for n, and suppose n^+ has a denumerable subset S. If $n \notin S$, then S is a denumerable subset of n (recall that "n" was defined to be the set $\{0, 1, \ldots, n - 1\}$); by the hypothesis of induction, this cannot happen. If $n \in S$, then $S - \{n\} \subseteq n$; but $S - \{n\}$ is denumerable (Lemma 7.13), so by the hypothesis of induction this cannot happen. We conclude that n^+ cannot have a denumerable subset. Now suppose that A has a denumerable subset B and A is finite; that is, $A \approx n$, $B \subseteq A$, and $B \approx \omega$. Then we have injective functions as follows: $\omega \to B \to A \to n$; their composite is an injective function $\omega \to n$, and we have just proven this to be impossible. Thus if A has a denumerable subset, then A is infinite. ■

7.15 Corollary Every set which has an infinite subset is infinite.

7.16 Corollary Every subset of a finite set is finite.

7.17 Corollary If A is an infinite set and B is nonempty, then $A \times B$ and $B \times A$ are infinite sets.

Proof. If y is a fixed element of B, the function $f: A \to A \times B$ given by $f(x) = (x, y)$ is clearly injective. Thus if $g: \omega \to A$ is a bijective function, then $f \circ g: \omega \to A \times B$ is injective. It follows that $A \times B$ has a denumerable subset, hence $A \times B$ is infinite. ■

7.18 Theorem A is an infinite set if and only if A is equipotent with a proper subset of itself.

---

* Note that, by definition, the natural number n is the set $\{0, 1, 2, \ldots, n - 1\}$, that is, n is exactly the initial segment S_n of ω.

Proof

i) Suppose that A is infinite; by Theorem 7.14, A has a denumerable subset $B = \{a_0, a_1, a_2, \ldots\}$. Let the function $f: A \to A$ be defined by

$$f(x) = x, \quad \forall x \in A - B,$$

$$f(a_m) = a_{m+1}, \quad \forall m \in \omega.$$

Clearly, f is a one-to-one correspondence between A and $A - \{a_0\}$.

ii) Suppose there exists a bijective function $f: A \to B$, where B is a proper subset of A. Let c be an arbitrary element of $A - B$; by the recursion theorem, there exists a function $\gamma: \omega \to A$ which satisfies the conditions

$$(\alpha)\ \gamma(0) = c \qquad \text{and} \qquad (\beta)\ \gamma(n^+) = f(\gamma(n)).$$

Now $\operatorname{ran} f = B$ and $c \in A - B$, so $c \notin \operatorname{ran} f$; thus by 6.9, γ is injective. The range of γ is obviously a denumerable subset of A, so by 7.14, A is infinite. ■

EXERCISES 7.3

1. Let A and B be a pair of disjoint finite sets. Use induction to prove that if $A \approx m$ and $B \approx n$, then $A \cup B \approx m + n$. Conclude that the union of two finite sets is finite.
2. Using the result of Exercise 1, prove that if A is an infinite set and B is a finite subset of A, then $A - B$ is infinite.
3. Prove that a natural number is not equipotent with a proper subset of itself. Conclude that if $A \approx m$ and $n > m$, then $A \not\approx n$.
4. Prove that A is an infinite set if and only if $\forall n \in \omega$, A has a subset B such that $B \approx n$.
5. Let A and B be finite sets. Use induction to prove that if $A \approx m$ and $B \approx n$, then $A \times B \approx mn$. Conclude that the Cartesian product of two finite sets is finite.
6. Assuming that if A is an infinite set and B is denumerable, prove that $A \approx (A \cup B)$.
7. Suppose $x \in A$; prove that A is an infinite set if and only if $A \approx (A - \{x\})$.
8. Use induction to prove that $A \approx n$, then $\mathscr{P}(A) \approx 2^n$. Conclude that if A is a finite set, then $\mathscr{P}(A)$ is a finite set.
9. Prove Corollary 7.15.
10. Prove Corollary 7.16.

4. PROPERTIES OF DENUMERABLE SETS

Once again, a set is called *denumerable* if it is in one-to-one correspondence with ω. The fundamental properties of denumerable sets are presented in this section.

7.19 Theorem Every subset of a denumerable set is finite or denumerable.

Proof. First we note that every subset of ω is finite or denumerable. Indeed, let $E \subseteq \omega$; by 4.63, either $E \simeq \omega$ or $E \simeq S_n = n$ for some $n \in \omega$. Now let A be a denumerable set and let $B \subseteq A$; there exists a bijective function $f: A \to \omega$. Now $f(B) \subseteq \omega$, so $f(B)$ is finite or denumerable; but f is bijective, so $B \approx f(B)$; hence B is finite or denumerable. ■

7.20 Theorem $\omega \times \omega \approx \omega$.

Proof. We will use the recursion theorem to establish the existence of a bijective function from ω to $\omega \times \omega$. Let $A = \omega \times \omega$; we define a function $f: A \to A$ as follows:

$$f(k, m) = \begin{cases} (0, k + 1) & \text{if } m = 0, \\ (k + 1, m - 1) & \text{if } m \neq 0. \end{cases}$$

We note that f is injective, for suppose

$$f(k, m) = f(n, p) = (r, s).$$

If $r = 0$, then (because of the way f is defined) $m = 0$ and $p = 0$; hence

$$(0, s) = f(k, m) = (0, k + 1) \quad \text{and} \quad (0, s) = f(n, p) = (0, n + 1),$$

so $k = n$. If $r \neq 0$, then

$$(r, s) = f(k, m) = (k + 1, m - 1) \quad \text{and} \quad (r, s) = f(n, p) = (n + 1, p - 1),$$

so $k = n$ and $m = p$. Thus f is injective. Now we make use of the recursion theorem: we define a function $\gamma: \omega \to A$ by the two conditions

i) $\gamma(0) = (0, 0)$, and
ii) $\gamma(n^+) = f(\gamma(n))$.

We note (again, because of the way f is defined) that $(0, 0)$ cannot be in the range of f; it follows by 6.9 that γ is injective. Finally, we show that γ is surjective; indeed, we will show that if $k, m \in \omega$, then $(k, m) = \gamma(n)$ for some $n \in \omega$. The proof is by induction on $k + m$:

I. If $k + m = 0$, then $(k, m) = (0, 0) = \gamma(0)$.

II. Suppose $k + m = n^+$; if $k = 0$, then $(k, m) = f(m - 1, 0)$; by the hypothesis of induction, $(m - 1, 0) = \gamma(q)$ for some $q \in \omega$, so

$$(k, m) = f(m - 1, 0) = f(\gamma(q)) = \gamma(q^+).$$

If $k \neq 0$, then $(k, m) = f(k - 1, m + 1)$; by the hypothesis of induction, $(k + m - 1, 0) = \gamma(p)$ for some $p \in \omega$; thus

$$\gamma(p + 1) = (0, k + m), \gamma(p + 2) = (1, k + m - 1), \ldots,$$
$$\gamma(p + k + 1) = (k, m). \blacksquare$$

7.21 Corollary If A and B are denumerable sets, then $A \times B$ is a denumerable set.

Proof. If $A \approx \omega$ and $B \approx \omega$, then $A \times B \approx \omega \times \omega$ (7.10). But $\omega \times \omega \approx \omega$; thus $A \times B \approx \omega$. ■

7.22 Theorem Let $\{A_n\}_{n\in\omega}$ be a denumerable family of denumerable sets, and let

$$A = \bigcup_{n\in\omega} A_n;$$

then A is denumerable. [A denumerable union of denumerable sets is denumerable.]

Proof. To say that each A_n is denumerable means that there exists a family $\{f_n\}_{n\in\omega}$ of functions such that, $\forall n \in \omega$, $f_n : \omega \to A_n$ is bijective. We define $\sigma : \omega \times \omega \to A$ by: $\sigma(k, m) = f_k(m)$. It is easy to see that σ is surjective: for if $x \in A$, then $x \in A_n$ for some $n \in \omega$, and if $x \in A_n$, then $x = f_n(m) = \sigma(n, m)$ for some $m \in \omega$.

By Theorem 7.20, there exists a bijective function $\phi : \omega \to \omega \times \omega$; hence $\sigma \circ \phi : \omega \to A$ is surjective. It follows (7.7) that $A \approx E$ for some subset $E \subseteq \omega$; now E is either finite or denumerable (Theorem 7.19); hence A is either finite or denumerable. By Corollary 7.15, A is not finite, so A is denumerable. ■

7.23 Corollary The union of two denumerable sets is denumerable.

EXERCISES 7.4

1. Prove that the union of two denumerable sets is denumerable. (Corollary 7.23.)
2. Let A be a denumerable set. Prove that A has a denumerable subset B such that $A - B$ is denumerable.
3. Prove that $\omega^n \approx \omega$. [*Hint*: Use the definitions $\omega^1 = \omega$ and $\omega^{n^+} = \omega^n \times \omega$; use 7.10, 7.20, and induction.] Conclude that if A is a denumerable set, then A^n is a denumerable set.
4. Prove that $\omega \cup \omega^2 \cup \omega^3 \cup \cdots$ is a denumerable set.

5. Prove that the set of all finite subsets of ω is denumerable. Then prove that the set of all finite subsets of a denumerable set is denumerable.

6. Let A be an infinite set. Prove that A is denumerable if and only if $A \approx B$ for every infinite subset $B \subseteq A$.

7. Prove that if A is a nonempty finite set and B is denumerable, then $A \times B$ is denumerable.

8. Let $\mathscr{L}$ be the set of all polynomials $a_0 + a_1x + \cdots + a_nx^n$ with integer coefficients. Prove that $\mathscr{L}$ is denumerable. [*Hint:* This may be proved by using an argument of the kind used in the introduction to prove that $\mathbb{Q}$ is denumerable.]

9. An *algebraic number* is any real root of an equation $a_0 + a_1x + \cdots + a_nx^n = 0$, where the coefficients a_i are integers. Prove that the set of all algebraic numbers is denumerable.

10. A real number is called *transcendental* if it is not algebraic. Prove that the set of all transcendental numbers is nondenumerable.

11. Use the results of Exercises 1 and 5, above, to prove that the set of all infinite subsets of ω is equipotent with 2^ω.

8

Arithmetic of Cardinal Numbers

1 INTRODUCTION

In the preceding chapter we defined what is meant by a finite set; it follows from our definition that *every finite set is equipotent with exactly one natural number n.* This fact has an important consequence, namely, that the natural numbers may be used as a set of standards—a scale, as it were—to measure the size of finite sets. If A is any finite set, then A may be "measured" by comparison with the natural numbers, and will be found to correspond—that is, to be equipotent—with exactly one of them. When the natural numbers serve in this capacity—as standards to measure the size of sets—they are commonly called *cardinal numbers.*

A natural and fascinating question arises now: Can we find a way of extending our system of cardinal numbers so as to create a set of standards for measuring the size of *all* sets? To put it another way: Can we define "infinite cardinal numbers," and can we construct a sufficient supply of them so that *every* set has a cardinal number (if A has n elements, we say that A *has cardinal number n*)? The answer is "yes:" We can generalize the concept of cardinal number with such remarkable ease that almost all of the properties of the finite cardinals—their ordering, their arithmetic, and so forth—apply as naturally to the infinite cardinals as they did to the finite ones. To take one example: What is meant by the sum $m + n$ of two cardinal numbers? The idea, clearly, is that if A has m elements and B has n elements—and if A and B are disjoint—then $m + n$ is the cardinal number of $A \cup B$. What could be more natural than to extend this notion of cardinal sum to all sets (or rather, to all "set sizes")?

Before giving a general definition of cardinal numbers, let us take one more look at the natural numbers and see why they can be used as standards to measure the size of finite sets. As we stated earlier, every finite set is equipotent with exactly one natural number n; that is, the natural numbers are well defined sets, and there is a unique natural number for each and every finite "set size." It is clear, now, what we expect of our definition of cardinal numbers: The cardinal numbers are to be *well-defined sets*, and *every set is to be equipotent with exactly one cardinal number.* It is immaterial what sets the cardinal numbers are; the only requirement is that there be exactly one cardinal number of each "size."

A simple way of constructing the cardinal numbers would be the following. We observe that the relation "A is equipotent with B" ($A \approx B$) is an equivalence relation among sets. Thus we might partition the class of all sets into equipotence classes, and select one representative of each class: the representatives would be our cardinal numbers. This process seems quite natural—and will, indeed, serve as the intuitive basis of our definition. However, it cannot be applied literally. Note, for example, that if A is a set, the equipotence class $\{B \mid B \approx A\}$ may be a proper class; hence it is not legitimate to speak of the "class of all the equipotence classes." Furthermore, even if we *could* speak of the "class of all the equipotence classes," it would not be legitimate to use the Axiom of Choice to pick a representative of each class; indeed, the Axiom of Choice (see statement Ch 1 on page 115) allows us to pick representatives from a *set of sets*, not from an arbitrary class of classes.

Since we cannot literally "select" our cardinals by using the Axiom of Choice, how are we to proceed? A simple way, and one which is sanctioned by mathematical tradition, is to *posit* their existence (that is, to posit the existence of a representative set from each "equipotence class") by means of a new axiom.

A10 Axiom of Cardinality There is a class CD of sets, called *cardinal numbers*, with the following properties:

K1 If A is any set, there exists a cardinal number a such that $A \approx a$.

K2 If A is a set and a, b are cardinal numbers, then $A \approx a$ and $A \approx b \Rightarrow a = b$.

We will add the Axiom of Cardinality to our list of axioms for set theory—but only on a provisional basis, for in the next chapter we will describe a method for *constructing* sets with properties K1 and K2—that is, we will produce actual sets (in much the same way as we produced the natural numbers) which will serve as cardinal numbers.

We will use lower-case Roman letters, such as a, b, c, d, etc., to denote cardinal numbers.

It is worth noting, incidentally, that *the class CD of all the cardinal numbers is a proper class.* For suppose CD is a set: since each cardinal number is a set, it follows by Axiom A5 that

$$V = \bigcup_{a \in CD} a$$

that is, the union of all the cardinal numbers, is a set. Thus by Axiom A6, $\mathscr{P}(V)$ is a set; but then, by condition K1 of Axiom A10, there is a cardinal number e such that $\mathrm{e} \approx \mathscr{P}(V)$. Now $\mathrm{e} \in CD$, hence $\mathrm{e} \subseteq V$, which is impossible by Corollary 7.2. This contradiction proves that CD is not a set, but a proper class.

2 OPERATIONS ON CARDINAL NUMBERS

If A is a set, a is a cardinal number and $A \approx a$, then we say that a is the cardinal number of A. We denote this by writing

$$a = \#A.$$

Now conditions K1 and K2 can be conveniently restated as follows:

K1 If A is any set, there exists a cardinal number a such that $a = \#A$.

K2 If A is a set and a, b are cardinal numbers, then $a = \#A$ and $b = \#A \Rightarrow a = b$.

8.1 Lemma If a and b are cardinal numbers and $a \approx b$, then $a = b$.

The proof is an immediate consequence of K2.

8.2 Lemma If $A \approx B$, then $\#A = \#B$.

Proof. By K1, there are cardinals a, b such that $a = \#A$ and $b = \#B$. Now $a \approx A$ and $b \approx B$; thus, if $A \approx B$, it follows that $a \approx A$ and $b \approx A$, so by K2, $a = b$. ■

We now proceed to define the addition and multiplication of cardinal numbers. Our definitions require no comment; they correspond in the most natural way to our intuitive understanding of the process of adding and multiplying whole numbers.

Let a and b be two cardinals. Let A and B be disjoint sets such that $a = \#A$ and $b = \#B$. Then $a + b$ is the cardinal number defined by

$$a + b = \#(A \cup B).$$

Note. In the preceding definition it has been assumed that we can always find *disjoint* sets A and B such that $a = \#A$ and $b = \#B$. This is obviously true. For example, take $A = a \times \{0\}$ and $B = b \times \{1\}$; then A consists of pairs $(x, 0)$, whereas B consists of pairs $(x, 1)$.

Let a and b be two cardinals. Let A and B be sets such that $a = \#A$ and $b = \#B$. Then ab is the cardinal number defined by

$$ab = \#(A \times B).$$

Note. Since a and b are sets, we can write $ab = \#(a \times b)$.

The usual algebraic laws for addition and multiplication follow from the elementary properties of sets.

8.3 Theorem If a, b, c are cardinal numbers, the following laws hold:

i) $a + b = b + a$,

ii) $ab = ba$,

iii) $a + (b + c) = (a + b) + c$,

iv) $a(bc) = (ab)c$,

v) $a(b + c) = ab + ac$.

Proof. Properties (i), (iii), and (v) are immediate consequences of Theorems 1.25(i), 1.25(v), and 1.32(ii) respectively. To prove (ii), it must be shown that there exists a one-to-one correspondence between $A \times B$ and $B \times A$; the function $\phi(x, y) = (y, x)$ is obviously such a correspondence. Finally, to prove (iv), we must show that there exists a one-to-one correspondence between $A \times (B \times C)$ and $(A \times B) \times C$; the function

$$\phi(x, (y, z)) = ((x, y), z)$$

is clearly such a correspondence. ■

Let A and B be finite sets. In Chapter 2 we defined A^B to be the set of all functions from B to A. Now suppose that A has m elements and B has n elements, and consider the process of constructing an arbitrary function from B to A. Since B has n elements, and each element can be assigned an image in one of m possible ways, this means there are exactly m^n distinct functions from B to A. This simple observation suggests the following definition of cardinal exponentiation:

Let a and b be two cardinals. Let A and B be sets such that $a = \#A$ and $b = \#B$. Then a^b is the cardinal number defined by

$$a^b = \#(A^B).$$

For notational convenience, we agree that $0^a = 0$ and $a^0 = 1$.

8.4 Theorem For any cardinals a, b, c the following rules hold:

i) $a^{b+c} = a^b a^c$,

ii) $(ab)^c = a^c b^c$,

iii) $(a^b)^c = a^{bc}$.

Proof. Let A, B, C be sets such that

$$a = \#A, \qquad b = \#B, \qquad c = \#C.$$

(For part (i) of the proof, assume $B \cap C = \varnothing$.)

i) We must show that there exists a bijective function $\sigma: A^{B\cup C} \to A^B \times A^C$.

We define σ as follows: If $f \in A^{B\cup C}$, then $\sigma(f) = (f_{[B]}, f_{[C]})$, where $f_{[B]}$ is the restriction of f to B and $f_{[C]}$ is the restriction of f to C. It is immediate that σ is a function from $A^{B\cup C}$ to $A^B \times A^C$.

σ is injective. Suppose $\sigma(f) = \sigma(g)$, where $f, g \in A^{B\cup C}$; then

$$(f_{[B]}, f_{[C]}) = (g_{[B]}, g_{[C]}),$$

that is,

$$f_{[B]} = f_{[C]} \qquad \text{and} \qquad g_{[B]} = g_{[C]};$$

Thus, by Theorem 2.15,

$$f = f_{[B]} \cup f_{[C]} = g_{[B]} \cup g_{[C]} = g.$$

σ is surjective. For if $(f_1, f_2) \in A^B \times A^C$ then, by Theorem 2.16,

$$f = f_1 \cup f_2 \in A^{B\cup C}, f_1 = f_{[B]}, \text{ and } f_2 = f_{[C]}; \text{ hence } (f_1, f_2) = \sigma(f).$$

ii) We will show that there exists a bijective function

$$\sigma: A^C \times B^C \to (A \times B)^C.$$

We define σ as follows: If $(f_1, f_2) \in A^C \times B^C$, then $\sigma(f_1, f_2)$ is the function f defined by

$$f(c) = (f_1(c), f_2(c)), \quad \forall c \in C;$$

certainly $f \in (A \times B)^C$. Now it is immediate that σ is a function from $A^C \times B^C \to (A \times B)^C$.

σ is injective. For if $f = \sigma(f_1, f_2) = \sigma(f_1', f_2') = f'$, then

$$\forall c \in C, \quad (f_1(c), f_2(c)) = f(c) = f'(c) = (f_1'(c), f_2'(c)),$$

so $f_1(c) = f_1'(c)$ and $f_2(c) = f_2'(c)$. It follows that $f_1 = f_1'$ and $f_2 = f_2'$, hence

$$(f_1, f_2) = (f_1', f_2').$$

σ is surjective. For if $f \in (A \times B)^C$, we may define

$$f_1 = \{(x, y) \mid (x, (y, z)) \in f\}$$

and

$$f_2 = \{(x, z) \mid (x, (y, z)) \in f\}.$$

It is easily shown that $f_1 \in A^C$, $f_2 \in B^C$, and $f = \sigma(f_1, f_2)$; the details are left to the reader.

iii) We will show that there exists a bijective function $\sigma: (A^B)^C \to A^{B\times C}$.

Note that if $f \in (A^B)^C$ and $c \in C$, then $f(c) \in A^B$; thus, if $b \in B$, then $[f(c)](b) \in A$. Now de define $\sigma(f)$ to be the function $\hat{f}$ given by

$$\hat{f}(b, c) = [f(c)](b).$$

Certainly, $\sigma(f) = \hat{f} \in A^{B \times C}$. Now it is immediate that σ is a function from $(A^B)^C$ to $A^{B \times C}$.

σ is injective. For if $\hat{f} = \sigma(f) = \sigma(f') = \hat{f}'$, then

$$\forall (b, c) \in B \times C, \quad [f(c)](b) = \hat{f}(b, c) = \hat{f}'(b, c) = [f'(c)](b).$$

Thus $\forall c \in C$, $f(c) = f'(c)$, so finally, $f = f'$.

σ is surjective. For if $g \in A^{B \times C}$ and $c \in C$, let f_c be defined by

$$f_c(b) = g(b, c), \quad \forall b \in B;$$

Clearly $f_c \in A^B$. Now, if f is given by $f(c) = f_c$, it is easily verified that f is a function from C to A^B; clearly $g = \sigma(f)$. ■

The finite cardinal numbers are designated, as usual, by the symbols 0, 1, 2, and so forth. It is to be especially noted that 0 is the cardinal number of the empty set, and 1 is the cardinal number of any singleton. The cardinal number of ω—that is, the cardinal number of any denumerable set—is customarily designated by the symbol $\aleph_0$ ("aleph-null").

It is useful to distinguish between *finite cardinal numbers*—that is, cardinal numbers of finite sets—and *infinite*, or *transfinite, cardinal numbers*, which are the cardinal numbers of infinite sets. We will see, shortly, that infinite cardinals have several properties which do not hold for finite cardinals.

EXERCISES 8.2

1. Prove each of the following, where a is any cardinal number.

   a) $a + 0 = a$, b) $a0 = 0$, c) $0^a = 0$.

2. Prove each of the following, where a is any cardinal number.

   a) $1a = a$, b) $a^1 = a$, c) $1^a = 1$.

3. If a, b are arbitrary cardinal numbers, prove that $ab = 0$ if and only if $a = 0$ or $b = 0$.

4. If a, b are arbitrary cardinal numbers, prove that $ab = 1$ if and only if $a = 1$ and $b = 1$.

5. Give a counterexample to the rule: $a + b = a + c \Rightarrow b = c$.

6. Give a counterexample to the rule: $ab = ac \Rightarrow b = c$.

7. If n is a finite cardinal number, use induction to prove that $na = a + a + \cdots + a$, where the right-hand side of the equality has n terms.

8. If n is a finite cardinal number, use induction to prove that $a^n = aa \cdots a$, where the right-hand side of the equality has n factors.

9. Let a, b be cardinals, and let A, B be sets such that $a = \#A$ and $b = \#B$. Prove that $a + b = \#(A \cup B) + \#(A \cap B)$.

10. Prove that if a is an infinite cardinal number and n is a finite cardinal number, then $a + n = a$.

11. Prove that if $a + 1 = a$, then a is an infinite cardinal number.

12. If b is an infinite cardinal number, prove that $\aleph_0 + b = b$.

3 ORDERING OF THE CARDINAL NUMBERS

Since cardinal numbers measure the size of sets, we naturally expect the cardinal number of a smaller set to be "less than" the cardinal number of a larger set. This suggests a natural ordering of the cardinal numbers:

Let a and b be cardinals, and let A and B be sets such that $a = \#A$ and $b = \#B$. The relation $\leqslant$ is defined by

$$a \leqslant b \qquad \text{if and only if} \qquad A \preccurlyeq B.$$

Note. Clearly, $a \leqslant b$ if and only if $a \preccurlyeq b$. In particular, $a \leqslant b$ if and only if there exists an injective function $f: a \to b$.

Our goal in this section is to show that the relation $\leqslant$ defined above is an order relation among the cardinal numbers, and, in particular, that the class of all the cardinal numbers is well ordered with respect to this relation.

8.5 Theorem (*Schröder-Bernstein*). Let a and b be cardinal numbers; if $a \leqslant b$ and $b \leqslant a$, then $a = b$.

Proof. Suppose $a \leqslant b$ and $b \leqslant a$; if A and B are sets such that $a = \#A$ and $b = \#B$, then $A \preccurlyeq B$ and $B \preccurlyeq A$, that is, there exist injective functions $f: A \to B$ and $g: B \to A$. If $C \subseteq A$, let $\Delta(C) = A - \bar{g}[B - \bar{f}(C)]$; it is easy to see that if C and D are subsets of A, then

1) $$C \subseteq D \text{ implies } \Delta(C) \subseteq \Delta(D).$$

Indeed,

$$C \subseteq D \Rightarrow \bar{f}(C) \subseteq \bar{f}(D) \qquad \text{(this is half of Theorem 2.29)}$$

$$\Rightarrow B - \bar{f}(D) \subseteq B - \bar{f}(C) \quad \text{by elementary class algebra}$$

$$\Rightarrow \bar{g}[B - \bar{f}(D)] \subseteq \bar{g}[B - \bar{f}(C)]$$
$$\Rightarrow A - \bar{g}[B - \bar{f}(C)] \subseteq A - \bar{g}[B - \bar{f}(D)].$$

Now, let $S = \{B \mid B \subseteq A \text{ and } B \subseteq \Delta(B)\}$, and let $A_1 = \bigcup_{B \in S} B$. We will prove that $A_1 = \Delta(A_1)$.

i) If $a \in A_1$, then $a \in B$ for some $B \in S$; but $B \subseteq A_1$, so by (1), $\Delta(B) \subseteq \Delta(A_1)$. Thus we have

$$a \in B \subseteq \Delta(B) \subseteq \Delta(A_1);$$

this proves that $A_1 \subseteq \Delta(A_1)$.

ii) We have just shown that $A_1 \subseteq \Delta(A_1)$, hence by (1), $\Delta(A_1) \subseteq \Delta[\Delta(A_1)]$, so $\Delta(A_1) \in S$. But A_1 is the union of all the elements of S, so $\Delta(A_1) \subseteq A_1$. Thus, we have proved that $A_1 = \Delta(A_1)$, which is the same as

$$A_1 = A - \bar{g}[B - \bar{f}(A_1)].$$

By elementary class algebra (see Exercise 11, Exercise Set 1.3) this gives

2) $$A - A_1 = \bar{g}[B - \bar{f}(A_1)].$$

Now, f and g are injective functions, hence $A_1 \approx \bar{f}(A_1)$ and, by (2),

$$B - \bar{f}(A_1) \approx \bar{g}[B - \bar{f}(A_1)] = A - A_1.$$

But $\bar{f}(A_1) \approx A_1$; thus, by 7.9, $A \approx B$. ■

It is immediate that the relation $\leqslant$ between cardinal numbers is reflexive and transitive; by 8.5 it is antisymmetric, hence it is an order relation.

8.6 Theorem Every class of cardinal numbers has a least element.

Proof. Let $\mathscr{A}$ be an arbitrary class of cardinal numbers, and let $a \in \mathscr{A}$; if a is the least element of $\mathscr{A}$, we are done; otherwise, let $\mathscr{B} = \{b \in \mathscr{A} \mid b < a\}$. Using the well-ordering theorem, let us well order a; for each $b \in \mathscr{B}$, let $\phi(b)$ be the least element $x \in a$ such that $b \approx S_x$. Now the set $\{\phi(b) \mid b \in \mathscr{B}\}$ has a least element $\phi(d)$ because it is a subset of a; we will show that d is the least element of $\mathscr{B}$. Indeed, let b be an arbitrary element of $\mathscr{B}$; $\phi(d) \leqslant \phi(b)$, hence $S_{\phi(d)} \subseteq S_{\phi(b)}$. Thus we have injective functions

$$d \to S_{\phi(d)} \xrightarrow{\lambda} S_{\phi(b)} \to b$$

(λ is the inclusion function), hence $d \leqslant b$. Thus d is the least element of $\mathscr{B}$, hence the least element of $\mathscr{A}$. ■

We are able to conclude:

8.7 The class of all the cardinal numbers, ordered by $\leqslant$, is well ordered.

The familiar "rules of inequality" apply to the cardinal numbers, as we shall see next.

8.8 Theorem Let a, b be cardinal numbers. Then $a \leqslant b$ if and only if there exists c such that $b = a + c$.

Proof

i) Suppose $b = a + c$; let A, B, C be sets (assume $A \cap C = \varnothing$) such that

$$a = \#A, \qquad b = \#B, \qquad \text{and} \qquad c = \#C.$$

Then there exists a bijective function $f: A \cup C \to B$. Clearly $f_{[A]}$ is an injective function from A to B, so $A \preccurlyeq B$.

ii) Suppose $a \leqslant b$; let A, B be disjoint sets such that

$$a = \#A, \qquad b = \#B.$$

There exists an injective function $f: A \to B$; since f is injective, $A \approx \overline{f}(A)$, so $a = \#f(A)$. If $C = B - f(A)$ and $c = \#C$, clearly $b = a + c$. ■

8.9 Theorem Let a, b, c, d be cardinal numbers. If $a \leqslant c$ and $b \leqslant d$, then we have the following:

i) $a + b \leqslant c + d$, ii) $ab \leqslant cd$, iii) $a^b \leqslant c^d$.

Proof. By Theorem 8.8, there exist r, s such that $c = a + r$ and $d = b + s$.

i) $c + d = a + r + b + s = (a + b) + (r + s)$, so by 8.8, $a + b \leqslant c + d$.

ii) $cd = (a + r)(b + s) = ab + as + rb + rs = ab + (as + rb + rs)$, so by Theorem 8.8, $ab \leqslant cd$.

iii) First we must show that $a^b \leqslant (a + r)^b$; that is, if A, R, B are sets such that $a = \#A$, $b = \#B$ and $r = \#R$, we must show that there exists an injective function $\sigma: A^B \to (A \cup R)^B$. We define σ by

$$\sigma(f) = f, \quad \forall f \in A^B,$$

and note (see 2.4) that a function $f: B \to A$ is also a function $f: B \to A \cup R$. It is immediate that σ is injective, hence $a^b \leqslant (a + r)^b$, that is $a^b \leqslant c^b$. Finally, using part (ii), we have

$$a^b = a^b 1 \leqslant c^b c^s = c^{b+s} = c^d. \ ■$$

8.10 *Remark.* It is important to note that 8.7 gives us valuable new information on the relation $A \preccurlyeq B$ among sets. Indeed, the following are two immediate consequences of 8.7:

1) If A and B are arbitrary sets, then $A \preccurlyeq B$ or $B \preccurlyeq A$.
2) If $A \preccurlyeq B$ and $B \preccurlyeq A$, then $A \approx B$.

Item (2) is especially useful when we need to prove that two sets are in one-to-one correspondence, for it is now sufficient to show that there is an injective function from A to B and an injective function from B to A (alternatively, a surjective function from A to B and a surjective function from B to A).

EXERCISES 8.3

1. Prove that if A and B are arbitrary sets, then $A \preccurlyeq B$ or $B \preccurlyeq A$.

2. Prove that if $A \preccurlyeq B$ and $B \preccurlyeq A$, then $A \approx B$.

3. Prove the following, where a, b, c, d are cardinal numbers.

   a) If $a^c < b^c$, then $a < b$.
   b) If $a^c < a^d$, then $c < d$.

4. Prove that if $a + b = c$, then $(r + s)^c \geqslant r^a s^b$.

5. Prove that there exists a strictly increasing sequence $a_1 < a_2 < \cdots$ of cardinal numbers, each with the property $a_i^{\aleph_0} = a_i$. [*Hint:* Take $a_1 = \aleph_0^{\aleph_0}$; then take $a_2 = 2^{a_1 \aleph_0}$, $a_3 = 2^{a_2 \aleph_0}$, etc.]

6. Let A be a denumerable set. Prove each of the following:

   a) $A^A \subseteq \mathscr{P}(A \times A)$. Conclude that $A^A \preccurlyeq \mathscr{P}(A \times A)$, hence $A^A \preccurlyeq \mathscr{P}(A)$.

   b) Verify that the function ϕ given by: $\phi(f) = \text{range } f, \forall f \in A^A$, is a surjective function $A^A \to \mathscr{P}(A)$. Conclude that $\mathscr{P}(A) \preccurlyeq A^A$.

   c) $A^A \approx \mathscr{P}(A)$; that is, $A^A \approx 2^A$. Conclude that $\aleph_0^{\aleph_0} = 2^{\aleph_0}$.

7. Use the argument outlined in the preceding problem to prove that the set of all injective functions $A \to A$ is equipotent with 2^A.

4 SPECIAL PROPERTIES OF INFINITE CARDINAL NUMBERS

A few remarkable arithmetic rules hold exclusively for *infinite* cardinals. As a result of these rules, the arithmetic of infinite cardinal numbers is a very simple matter.

8.11 Theorem If a is an infinite cardinal number, then $aa = a$.

Proof. Let A be a set such that $a = \#A$. Since A is infinite, A has a denumerable subset D. By Corollary 7.21, $D \approx D \times D$; that is, there exists a bijective function $\phi: D \to D \times D$. Now let $\mathscr{A}$ be the set of all pairs $(B. f)$ which satisfy the following conditions:

i) B is a subset of A and f is a bijective function from B to $B \times B$.
ii) $D \subseteq B$.
iii) $\phi \subseteq f$.

We order $\mathscr{A}$ by the relation $(B_1, f_1) \leqslant (B_2, f_2)$ iff $B_1 \subseteq B_2$ and $f_1 \subseteq f_2$. $\mathscr{A}$ is nonempty, for $(D, \phi) \in \mathscr{A}$. Now it is easy to verify that $\mathscr{A}$ satisfies the hypotheses of Zorn's Lemma (the details are left as an exercise for the reader). Thus $\mathscr{A}$ has a maximal element (C, g); it remains only to show that $\#C = a$. We will prove this by contradiction—assuming that $\#C < a$ and proving this to be impossible.

Let $b = \#C$ and assume that $b < a$. Since $C \times C \approx C$, it follows that $bb = b$; furthermore,

$$b = 0 + b \leqslant b + b$$

and

$$b + b = 1b + 1b = (1 + 1)b \leqslant bb = b;$$

hence $b = b + b$. Now let $d = \#(A - C)$; C and $A - C$ are disjoint, so

$$a = \#A = \#(A - C) + \#C = d + b.$$

We note that $b < d$, for $d \leqslant b$ implies that

$$a = d + b \leqslant b + b = b,$$

which would contradict our assumption that $b < a$. From $b < d$ it follows that $A - C$ has a subset E such that $\#E = b$.

Now

$$(C \cup E) \times (C \cup E) = (C \times C) \cup (C \times E) \cup (E \times C) \cup (E \times E),$$

where $C \times C, C \times E, E \times C, E \times E$ are mutually disjoint sets, each of which has the cardinal $bb = b$. Thus

$$\#[(C \times E) \cup (E \times C) \cup (E \times E)] = b + b + b = (b + b) + b = b + b = b,$$

hence there exists a bijective function

$$h: E \to [(C \times E) \cup (E \times C) \cup (E \times E)].$$

It follows by 7.8 that $g \cup h$ is a bijective function from $C \cup E$ to

$$(C \times C) \cup [(C \times E) \cup (E \times C) \cup (E \times E)] = (C \cup E) \times (C \cup E),$$

hence $(C \cup E, g \cup h) > (C, g)$, which is impossible because (C, g) is a maximal element of $\mathscr{A}$.

The assumption that $b < a$ has led to a contradiction; thus $b = a$, so $aa = a$. ■

8.12 Corollary Let a and b be cardinals, where a is infinite and $b \neq 0$. If $b \leqslant a$, then $ab = a$.

Proof. Since $b \geqslant 1$, thus $a = a1 \leqslant ab$; but $ab \leqslant aa = a$, hence $ab = a$. ■

8.13 Corollary If a is an infinite cardinal, $a + a = a$.

Proof. We have $a = 1a \leqslant 2a \leqslant aa = a$; but $2a = (1 + 1)a = a + a$, so $a + a = a$. ■

8.14 Corollary Let a and b be cardinals, where a is infinite. If $b \leqslant a$, then $a + b = a$.

Proof. We have $a = a + 0 \leqslant a + b$; but $a + b \leqslant a + a = a$, so $a + b = a$. ■

8.15 Corollary Let a and b be infinite cardinal numbers. Then

$$a + b = ab = \max\{a, b\}.$$

8.16 Theorem Let $a > 1$ be a cardinal number and let b be an infinite cardinal number. If $a \leqslant b$, then $a^b = 2^b$.

Proof. By 7.4, $a < 2^a$, so $a^b \leqslant (2^a)^b = 2^{ab}$. But by Corollary 8.12 $ab = b$, so $a^b \leqslant 2^b$. On the other hand, $2 \leqslant a$, so $2^b \leqslant a^b$. Consequently $a^b = 2^b$. ■

Remark. Theorem 8.11 and its corollaries can be interpreted very profitably in terms of sets and the relation $A \preccurlyeq B$ among sets. For example, Theorem 8.11 tells us that if A is an infinite set, then $A \approx A \times A$. This has an interesting consequence: $A \times A$ has a partition $\{B_x\}_{x \in A}$ where $B_x = \{(x, y) \mid y \in A\}$. Hence the bijective function from $A \times A$ to A induces a corresponding partition $\{C_x\}_{x \in A}$ of A, where A is the index set and each member of the partition is equipotent with A.

EXERCISES 8.4

1. If a is an infinite cardinal number and $a \leqslant bc$, prove that $a \leqslant b$ or $a \leqslant c$.
2. Let a be a cardinal number > 1, and let b be an infinite cardinal number. Prove that if $a = a^b$, then $b < a$.

3. An infinite cardinal number a is said to be *dominant* if it satisfies the following condition: if b and c are cardinal numbers such that $b < a$ and $c < a$, then $b^c < a$. Prove that a is a dominant cardinal number if and only if $d < a \Rightarrow 2^d < a$.

4. If a, c, and d are arbitrary cardinal numbers and b is an infinite cardinal number, prove that

$$a + b \leqslant c + d \Rightarrow a \leqslant c \text{ or } b \leqslant d.$$

5. Let a, b, c, d be cardinal numbers. Prove that if $a < b$ and $c < d$, then $ac < bd$ and $a + c < b + d$. [*Hint:* For the case where b and d are both finite, this result has been proven in Exercises 5 and 6, Exercise Set 6.4. Ignore this case, and assume that b is infinite or d is infinite (this assumption includes three cases).]

In Exercises 6 through 10, a, b, and c are arbitrary cardinals. For each of these problems, the case where a, b, and c are all finite has been considered in Chapter 6. Ignore this case, and treat the remaining cases.

6. Prove that if $a + a = a + b$, then $a \geqslant b$.

7. Prove that if $a + b < a + c$, then $b < c$.

8. Prove that if $ab < ac$, then $b < c$.

5 INFINITE SUMS AND PRODUCTS OF CARDINAL NUMBERS

Early in this book we spoke of the union of two classes; later we extended this notion by defining the union of an arbitrary family of classes. Similarly, we introduced the Cartesian product of two classes and later generalized this to the product of a family of classes. In both cases, extending our original definition seemed like a perfectly natural thing to do, for the intuitive concepts of union and product can be applied as easily to a family of classes as to a pair of classes. The same holds true for the process of adding and multiplying cardinal numbers; they lend themselves to the following obvious generalization.

Let $\{a_i\}_{i\in I}$ be a family of cardinal numbers; let $\{A_i\}_{i\in I}$ be a family of disjoint sets such that $a_i = \#A_i$ for each $i \in I$. Then $\sum_{i\in I} a_i$ is the cardinal number defined by

$$\sum_{i\in I} a_i = \#(\bigcup_{i\in I} A_i).$$

Let $\{a_i\}_{i\in I}$ be a family of cardinal numbers, and let $\{A_i\}_{i\in I}$ be a family of sets such that $a_i = \#A_i$ for each $i \in I$. Then $\bigtimes_{i\in I} a_i$ is the cardinal number defined by

$$\bigtimes_{i\in I} a_i = \#(\prod_{i\in I} A_i).$$

In elementary arithmetic we learn that ab is the result of "adding a to itself b times," and that a^b is the result of "multiplying a by itself b times." It is useful to know that this holds true for all cardinal numbers a and b.

8.17 Theorem Let a and b be cardinal numbers, and let I be a set such that $b = \#I$. If $a = a_i$, $\forall i \in I$, then

i) $ab = \sum_{i \in I} a_i$, and

ii) $a^b = \bigtimes_{i \in I} a_i$.

Proof

i) Let $\{A_i\}_{i \in I}$ be a family of disjoint sets such that $a = a_i = \#A_i$ for each $i \in I$, and let A be a set such that $a = \#A$. Since $A_i \approx A$ for each $i \in I$, there exists a family $\{f_i: A \to A_i\}_{i \in I}$ of bijective functions. We define

$$f: A \times I \to (\bigcup_{i \in I} A_i)$$

by

$$f(x, i) = f_i(x);$$

it is elementary to verify that f is bijective. Thus

$$A \times I \approx (\bigcup_{i \in I} A_i);$$

that is,

$$ab = \sum_{i \in I} a_i.$$

ii) We wish to show that $A^I \approx \prod_{i \in I} A_i$, where $A_i = A$ for each $i \in I$. But a glance at the definitions of A^I and $\prod_{i \in I} A_i$ (where $A_i = A$, $\forall i \in I$) will reveal that they both refer to the same set—the set of all functions from I to A. ■

Theorem 8.9 has the following analogue for infinite sums and products.

8.18 Theorem Let $\{a_i\}_{i \in I}$ and $\{b_i\}_{i \in I}$ be families of cardinal numbers. If $a_i \leqslant b_i$ for each $i \in I$, then

i) $\sum_{i \in I} a_i \leqslant \sum_{i \in I} b_i$,

ii) $\bigtimes_{i \in I} a_i \leqslant \bigtimes_{i \in I} b_i$.

Proof

i) Let $\{A_i\}_{i \in I}$ and $\{B_i\}_{i \in I}$ be families of disjoint sets such that $a_i = \#A_i$ and $b_i = \#B_i$ for each $i \in I$. Since $a_i \leqslant b_i$ for every $i \in I$, there exists a family $\{f_i: A_i \to B_i\}_{i \in I}$ of injective functions. It is easy to verify that $f = \bigcup_{i \in I} f_i$ is an injective function from $\bigcup_{i \in I} A_i$ to $\bigcup_{i \in I} B_i$. (The details are left as an exercise for the reader.)

ii) Given the family $\{f_i : A_i \to B_i\}_{i \in I}$ introduced above, we define a function

$$f : \prod_{i \in I} A_i \to \prod_{i \in I} B_i$$

as follows: if $x \in \prod_{i \in I} A_i$ and $y \in \prod_{i \in I} B_i$, then

$$f(x) = y \quad \text{iff} \quad f_i(x_i) = y_i \quad \forall i \in I.$$

We verify that f is injective: If $f(u) = f(v)$, then

$$f_i(u_i) = f_i(v_i), \quad \forall i \in I.$$

But each f_i is injective, so $u_i = v_i$ for every $i \in I$; hence $u = v$. ■

Theorems 8.17 and 8.18 have the following useful corollary.

8.19 Corollary Let $\{a_i \mid i \in I\}$ be a set of cardinal numbers, and let b and c be cardinal numbers. If $a_i \leqslant b$ for each $i \in I$ and if $\#I = c$, then

i) $\sum_{i \in I} a_i \leqslant bc$, and

ii) $\bigtimes_{i \in I} a_i \leqslant b^c$.

The proof, which follows immediately from 8.17 and 8.18, is left as an exercise for the reader.

EXERCISES 8.5

1. Prove that $\bigtimes_{i \in I} a_i = 0$ if and only if $a_i = 0$ for some $i \in I$.
2. Suppose $a_i \leqslant a$, $\forall i \in I$, and $\#I \leqslant a$, where a is some fixed cardinal. Prove that $\sum_{i \in I} a_i \leqslant a$. [*Hint:* Use Theorems 8.17 and 8.18.]
3. Suppose $a_i \leqslant a$, $\forall i \in I$, and $\#I \leqslant a$. Prove that $\bigtimes_{i \in I} a_i \leqslant 2^a$.
4. Let $\{a_i\}_{i \in I}$ be a set of cardinal numbers, and suppose there is no greatest element in this set. Prove that $\forall j \in I$, $a_j < \sum_{i \in I} a_i$.
5. Prove that $a \cdot \sum_{i \in I} b_i = \sum_{i \in I} ab_i$.
6. Use Theorem 8.18 to justify each of the following. (Each sum is understood to have $\aleph_0$ terms.)

   a) $1 + 2 + 3 + \cdots = \aleph_0$, b) $n + n + \cdots = \aleph_0$, c) $\aleph_0 + \aleph_0 + \cdots = \aleph_0$.

7. Let $f: A \to B$ be a surjective function, where B is an infinite set. If, $\forall y \in B, f^{-1}(y)$ is finite or denumerable, prove that $A \approx B$.

8. Let A be an infinite set, and let $F(A)$ designate the family of all finite subsets of A. Prove that $F(A) \approx A$. [*Hint:* For each $n \in \omega$, let F_n designate the family of n-element subsets of A. There exists an obvious surjective function from A^n to F_n; there are $\aleph_0$ sets F_n.]

9. If $\{C_i\}_{i \in I}$ is a family of sets, prove that $\#(\bigcup_{i \in I} C_i) \leqslant \sum_{i \in I} \#C_i$.

10. Let $\{a_i\}_{i \in I}$ and $\{a_i\}_{i \in J}$ be families of cardinal numbers. Prove that $\sum_{i \in I} a_i \leqslant \sum_{i \in I \cup J} a_i$.

9
Arithmetic of the Ordinal Numbers

1 INTRODUCTION

In elementary school we learn that there are cardinal numbers and ordinal numbers. The cardinal numbers, we are told, are the "counting" numbers: 1, 2, 3, and so on; the ordinal numbers are the "ranking" numbers: first, second, third, etc. The distinction may appear to be somewhat pedantic, for the natural numbers serve in both capacities, as ordinals and as cardinals, and there is no need in elementary arithmetic to differentiate between the two. However, one of the unexpected discoveries of modern set theory is that, just as the infinite cardinals behave differently from the finite ones, so the infinite ordinals exhibit a strikingly different behaviour from the cardinals. It is the purpose of this chapter to introduce the ordinal numbers and explore their properties—especially those of the infinite ordinals.

The dichotomy between cardinal and ordinal, from the scholastic point of view, arises from two different ways of *using* the natural numbers. In their role as cardinals, the natural numbers measure the "size," or power, of sets; in their role as ordinals, they serve to designate the rank, or position, of an object in a linearly ordered array. We will use this insight—although it is somewhat outdated—as the starting point of our discussion.

When we speak of ranking elements in some order, what kind of order do we have in mind? There must be a first element, a second element, and so on—in other words, the order is that of the natural numbers, which is a *well-ordering.* Now the reader should note that the general notion of well-ordering is an extension of the order of the natural numbers. Every infinite well-ordered set has a first element, a second element, and—for each natural number n—an nth element; but it may also have elements which are "beyond the reach" of the finite ordinals. Thus the set

$$\{x_1, x_2, x_3, \ldots, y_1, y_2, y_3, \ldots\}$$

has a first element x_1, a second element x_2, and so forth; but y_1, for example, though it has a perfectly well-defined "position" in the set, cannot be described as the "nth element" for any finite n.

Situations of this kind arise frequently in almost every branch of mathematics. For example, on page 141 we defined a sequence of sets by these conditions: $\omega = K_1$, $\mathscr{P}(K_1) = K_2$, and so on; $\bigcup_{i \in \omega} K_i = L_1$, $\mathscr{P}(L_1) = L_2$, etc. Continuing in this manner, we get the well-ordered family of sets

$$\{K_1 \prec K_2 \prec K_3 \prec \cdots \prec L_1 \prec L_2 \prec L_3 \prec \cdots \prec M_1 \prec M_2 \prec M_3 \prec \cdots\}.$$

Now K_1 is the first element of this family, and, in general, K_n is the nth element; but what of (say) L_1? Its position in the family is unambiguous: L_1 immediately follows *all* of the sets K_i; yet classical mathematics has not provided us with any ordinal number o which would enable us to describe L_1 as the "oth element of the set."

Thus, as in our study of cardinal numbers, we are led to ask an intriguing question: Can we find a way of extending our system of ordinal numbers so as to create a set of standards for designating the position of any element in any well-ordered set? The answer, once again, is "yes;" we can generalize the concept of ordinal number with such remarkable ease that no barrier, either logical or intuitive, seems to separate the finite ordinals from the infinite ones.

We will approach the ordinals in much the same way that we approached the cardinals. We will begin by defining a relation of "having the same ordinal number," and later define the ordinals, essentially, to be representative of the distinct classes induced by this relation.

If A and B are well-ordered sets, and if $x \in A$ and $y \in B$, then to say that "x has the same rank as y" (for example, x is the third element of A and y is the third element of B) is the same as saying that the initial segment S_x is isomorphic with the initial segment S_y. To say that "x has a lower rank than y" is the same as saying that S_x is isomorphic with an initial segment of S_y. The reader should stop here until he has thoroughly understood this fact, for it is the point of departure for achieving an understanding of the modern approach to the ordinal numbers. Isomorphism plays the same role in the study of ordinal numbers that one-to-one correspondence plays in the study of cardinal numbers.

An important warning needs to be given here. The alert reader may question the necessity of introducing the concept of isomorphism. After all, he may ask, why not say that x and y have the same rank if and only if S_x is *equipotent* with S_y? Surely if ten elements precede x and ten elements precede y, then x and y are both eleventh in their class. This is true when we are dealing with finite rank, but untrue in the case of infinite rank; a simple example will convince the reader. In the set

$$\{x_1, x_2, x_3, \ldots, y_1, y_2, y_3, \ldots\},$$

both y_1 and y_2 are preceded by a denumerable number of elements—that is, S_{y_1} is equipotent with S_{y_2}—yet y_2 clearly follows y_1.

When we say that x has the same rank as y, or x has a lower rank than y, we are only *apparently* speaking of x and y; actually, we are comparing the initial segments S_x and S_y. Hence we lose nothing if we confine our attention to the study of initial segments of well-ordered sets. But we can go a step further: An initial segment (of a well-ordered set) is a well-ordered set, and conversely, every well-ordered set A is an initial segment (adjoin a last element x to A—then A is S_x). Hence the study of ordinality is, essentially, the study of well-ordered sets.

Motivated by the foregoing remarks, we introduce the following definitions:

Let A and B be well-ordered sets. We say that A and B are *similar* (or have the *same ordinality*) if A is isomorphic with B; we write $A \approxeq B$. If A is isomorphic with an initial segment of B, we say that B is a *continuation* of A, or A has a *lower ordinality* than B, and we write $A \prec B$.

It follows from Theorem 4.62 that if A and B are any two well-ordered sets, then $A \approxeq B$, or $A \prec B$, or $B \prec A$.

In conclusion, to say that x has the same rank as y is the same as saying that $S_x \approxeq S_y$, and to say that x has a lower rank than y is the same as saying that $S_x \prec S_y$. Thus we have completely captured—and formalized—the intuitive concept of "rank," and have extended it beyond the unnatural confines of finite ordinality.

As for the ordinal numbers, we simply imitate the procedure we followed for the cardinals by introducing the

A11 Axiom of Ordinality There is a class OR of well-ordered sets, called *ordinal numbers*, with the following properties:

O1 If A is any well-ordered set, there exists an ordinal number α such that $A \approxeq \alpha$.

O2 If A is a well-ordered set and α, β are ordinal numbers, then $A \approxeq \alpha$ and $A \approxeq \beta \Rightarrow \alpha = \beta$.

We will add the Axiom of Ordinality to our list of axioms for set theory—but only on a provisional basis, for in the last section of this chapter we will describe a method for *constructing* sets with properties 01 and 02; those sets will then serve the purpose of ordinal numbers.

It is worth noting, incidentally, that the *class of all the ordinal numbers* is a proper class. Indeed, let OR be the class of all the ordinal numbers, and suppose OR is a set; from this assumption we will derive a contradiction. Let $A = \mathscr{P}(B)$, where $B = \bigcup_{\in \mathrm{OR}} \alpha$; since each ordinal number α is a set and OR (under our assumption) is a set, it follows by Axioms A5 and A6 that B, and therefore A, are sets. Let us well-order A; by 01, there is an ordinal number

α such that $\alpha \approxeq A$. But $\alpha \in \text{OR}$, hence $\alpha \subseteq B$, so by 7.2, α cannot be equipotent with $A = \mathscr{P}(B)$. This contradiction proves that OR is a proper class.

2 OPERATIONS ON ORDINAL NUMBERS

Following our "naive" introduction to ordinal numbers in the preceding section, we now proceed to study the ordinals from a formal point of view. We will henceforth consider the ordinals to be the objects defined by Conditions 01 and 02. The reader should adjust his thinking accordingly; he should cease thinking of ordinals as "symbols for designating rank," and begin to think of them as certain *well-ordered sets.*

9.1 Definition If A is a well-ordered set, α is an ordinal number, and $A \approxeq \alpha$, then we say that *α is the ordinal number of A.* We denote this by writing

$$\alpha = \circledcirc A.$$

Now Conditions 01 and 02 can be conveniently restated as follows:

O1 If A is a well-ordered set, there exists an ordinal number α such that $\alpha = \circledcirc A$.

O2 If A is a well-ordered set and α, β are ordinal numbers, then $\alpha = \circledcirc A$ and $\beta = \circledcirc A \Rightarrow \alpha = \beta$.

9.2 Lemma If α and β are ordinal numbers and $\alpha \approxeq \beta$, then $\alpha = \beta$.

The proof is an immediate consequence of 02.

9.3 Lemma If $A \approxeq B$, then $\circledcirc A = \circledcirc B$.

The proof is analogous to that of Lemma 8.2.

Before defining the addition and multiplication of ordinal numbers, we need to introduce two new operations on well-ordered sets.

9.4 Definition Let A and B be disjoint, well-ordered sets. $A \oplus B$, called the *ordinal sum* of A and B, is the set $A \cup B$ ordered as follows. If $x, y \in A \cup B$, then $x \leqslant y$ if and only if

i) $x \in A$ and $y \in A$ and $x \leqslant y$ in A, or
ii) $x \in B$ and $y \in B$ and $x \leqslant y$ in B, or
iii) $x \in A$ and $y \in B$.

Thus, in $A \oplus B$, the elements of A are ordered as before, the elements of B are ordered as before, and every element of B is greater than every element of A.

Having defined the ordinal sum of two well-ordered sets, it is natural to define the ordinal sum of an arbitrary family of well-ordered sets.

9.5 Definition Let I be a set, let $\{A_i\}_{i\in I}$ be a family of disjoint well-ordered sets and let the index set I be well-ordered.

$\mathbf{S}_{i\in I} A_i$, called the *ordinal sum* of the family $\{A_i\}_{i\in I}$, is the set $\bigcup_{i\in I} A_i$ ordered in the following way: if $x, y \in \bigcup_{i\in I} A_i$, then $x \leqslant y$ if and only if

i) for some $i \in I$, $x \in A_i$ and $y \in A_i$ and $x \leqslant y$ in A_i, or
ii) $x \in A_i$ and $y \in A_j$ and $i < j$.

Thus, in $\mathbf{S}_{i\in I} A_i$, each set A_i is ordered as before, and, for $i < j$, every element of A_i is less than every element of A_j.

An easy step leads us, now, to the notion of ordinal product. To put it simply, the product $A \odot B$ is the result of "adding A to itself B times." More precisely, if $\{A_i\}_{i\in B}$ is a family of disjoint, well-ordered sets, indexed by B, where each A_i is similar to A, then $A \odot B$ is the set $\mathbf{S}_{i\in B} A_i$. The only remaining difficulty is to produce the family $\{A_i\}_{i\in B}$. To do so is easy enough: for each $i \in B$, we define A_i to be the set $\{(x, i) \mid x \in A\}$—that is, $A_i = A \times \{i\}$. But a happy thought strikes us now, as we realize that the set $\mathbf{S}_{i\in B} A_i$ is none other than the Cartesian product $A \times B$ ordered by the antilexicographic ordering (Definition 4.2). We exploit this fortunate coincidence to give the following elegant definition of ordinal product:

9.6 Definition Let A and B be well-ordered sets. Then $A \odot B$, called the *ordinal product* of A and B, is the set $A \times B$ ordered by the antilexicographic ordering.

9.7 Example Let $A = \{a, b, c\}$ be well ordered as follows: $a < b < c$. Let $B = \{1, 2, 3\}$ be well ordered as follows: $1 < 2 < 3$. Then $A \odot B$ is the set $A \times B$ well ordered as follows (see Fig. 9–1):

$$(a, 1) < (b, 1) < (c, 1) < (a, 2) < (b, 2) < (c, 2) < (a, 3) < (b, 3) < (c, 3).$$

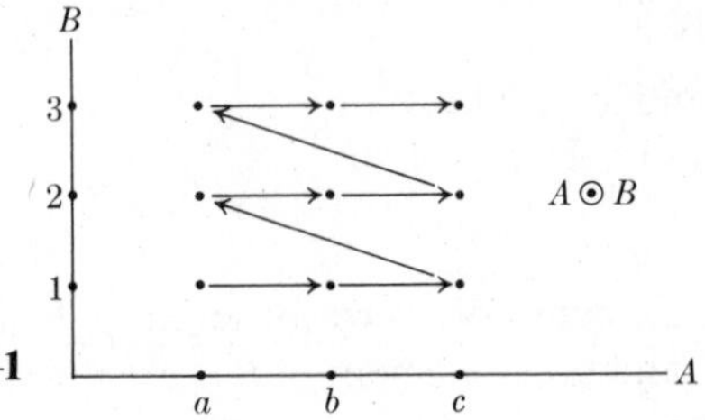

Fig. 9–1

Now, back to the ordinal numbers.

9.8 Definition Let α and β be ordinal numbers, and let A and B be disjoint, well-ordered sets such that $\alpha = \obslash A$ and $\beta = \obslash B$. We define the sum $\alpha + \beta$ to be the ordinal number given by

$$\alpha + \beta = \obslash(A \oplus B).$$

Let α and β be ordinal numbers, and let A and B be well-ordered sets such that $\alpha = \obslash A$ and $\beta = \obslash B$. We define the product $\alpha\beta$ to be the ordinal number given by

$$\alpha\beta = \obslash(A \odot B).$$

The elementary properties of ordinal addition and multiplication are given in the following theorem.

9.9 Theorem Let α, β and γ be ordinal numbers. Then

i) $\alpha + (\beta + \gamma) = (\alpha + \beta) + \gamma$,
ii) $\alpha(\beta\gamma) = (\alpha\beta)\gamma$,
iii) $\alpha(\beta + \gamma) = \alpha\beta + \alpha\gamma$.

Proof. Let A, B, and C be disjoint, well-ordered sets such that $\alpha = \obslash A$, $\beta = \obslash B$, and $\gamma = \obslash C$.

i) We must show that

$$A \oplus (B \oplus C) \approxeq (A \oplus B) \oplus C.$$

But it follows immediately from our definition of ordinal sums that both $A \oplus (B \oplus C)$ and $(A \oplus B) \oplus C$ designate the set $A \cup B \cup C$ with the following order: If x and y are both in A, both in B, or both in C, they are ordered according to their order in A, B, or C respectively; furthermore, every element of C is greater than every element of B, and every element of B is greater than every element of A.

ii) We must show that

$$A \odot (B \odot C) \approxeq (A \odot B) \odot C.$$

We have seen earlier that the function

$$f\big(x, (y, z)\big) = \big((x, y), z\big)$$

is a one-to-one correspondence between $A \times (B \times C)$ and $(A \times B) \times C$. In order to establish that f is an isomorphism, we need simply show that

$$\big(x, (y, z)\big) \leqslant \big(x', (y', z')\big)$$

if and only if

$$((x, y), z) \leqslant ((x', y'), z').$$

The details, which follow immediately from the definition of ordinal product, are left as an exercise for the reader.

iii) We must show that

$$A \odot (B \oplus C) \cong (A \odot B) \oplus (A \odot C).$$

Both $A \odot (B \oplus C)$ and $(A \odot B) \oplus (A \odot C)$ designate the same set,

$$A \times (B \cup C) = (A \times B) \cup (A \times C),$$

with certain orderings; it is easy to show that the two orderings are the same. The details are left as an exercise for the reader. ■

As usual, 0 is the ordinal number of the empty set, and 1 is the ordinal number of any singleton. An ordinal number μ is said to be *finite* if μ is similar to a natural number n; if μ is not a finite ordinal, then μ is called an *infinite*, or *transfinite*, ordinal. It is customary to designate the ordinal number of ω by means of the symbol ω.

It is most important to note that *addition and multiplication of ordinal numbers are not commutative.* Two simple examples will suffice to establish this fact. First, let us take addition. If we compare $\omega + 1$ with $1 + \omega$, we observe that $1 + \omega$ is similar to ω, whereas $\omega + 1$ is not (it has a last element!); thus $\omega + 1 \neq 1 + \omega$. Next, let us take multiplication. We observe that

$$2\omega = \{(0, 0), (1, 0), (0, 1), (1, 1), (0, 2), (1, 2), \ldots\}$$

and that

$$\omega 2 = \{(0, 0), (1, 0), (2, 0), \ldots, (0, 1), (1, 1), (2, 1), \ldots\};$$

these sets are obviously not isomorphic, hence $\omega 2 \neq 2\omega$.

We note also that the "*right distributive law*" $(\alpha + \beta)\gamma = \alpha\gamma + \beta\gamma$ *does not hold.* For example,

$$(1 + 1)\omega = 2\omega,$$

whereas

$$1\omega + 1\omega = \omega + \omega = \omega(1 + 1) = \omega 2,$$

and we noted in the preceding paragraph that $2\omega \neq \omega 2$; thus

$$(1 + 1)\omega \neq 1\omega + 1\omega.$$

EXERCISES 9.2

1. Let A_1, A_2, B_1, B_2 be well-ordered sets. Prove that if $A_1 \approxeq A_2$ and $B_1 \approxeq B_2$, then
   a) $A_1 \oplus B_1 \approxeq A_2 \oplus B_2$, and
   b) $A_1 \odot B_1 \approxeq A_2 \odot B_2$.
2. If A and B are well-ordered sets, prove that $A \oplus B$ and $A \odot B$ are well-ordered sets.
3. If α is an ordinal number, prove that $1 + \alpha = \alpha$ iff α is an infinite ordinal.
4. Let α and β be nonzero ordinal numbers. Prove that if $\alpha + \beta = \omega$, then α is a finite ordinal number (that is, similar to a natural number) and $\beta = \omega$. Now assume $\beta \neq 1$. Prove that if $\alpha\beta = \omega$, then α is finite and $\beta = \omega$. [*Hint:* Consider the well-ordered sets $A \oplus B$ and $A \odot B$, where $\alpha = \circledcirc A$ and $\beta = \circledcirc B$.]
5. Prove each of the following, where μ designates a finite ordinal.
   a) $\mu + \omega = \omega$,
   b) $\mu\omega = \omega$,
   c) If α is an infinite ordinal, then $\mu + \alpha = \alpha$.
6. Prove that $(\omega + \omega)\omega = \omega\omega$.
7. Give a counterexample to the (false) rule
$$\alpha + \gamma = \beta + \gamma \Rightarrow \alpha = \beta.$$
8. Prove the following, for every ordinal number α.
   a) $0 + \alpha = \alpha + 0 = \alpha$, b) $\alpha 0 = 0\alpha = 0$, c) $\alpha 1 = 1\alpha = \alpha$.
9. Prove that $\alpha\beta = 0$ if and only if $\alpha = 0$ or $\beta = 0$.
10. Prove each of the following, where μ, ν, π are finite ordinals.
    a) $m \approxeq \mu$ iff $m \approx \mu$ (where $m \in \omega$), b) $\mu + \nu = \nu + \mu$,
    c) $\mu\nu = \nu\mu$, d) $(\mu + \nu)\pi = \mu\pi + \nu\pi$.
11. Give a complete proof of the isomorphism [Theorem 9.9(ii)]
$$A \odot (B \odot C) \approxeq (A \odot B) \odot C.$$
12. Give a complete proof of the isomorphism [Theorem 9.9(iii)]
$$A \odot (B \oplus C) \approxeq (A \odot B) \oplus (A \odot C).$$

3 ORDERING OF THE ORDINAL NUMBERS

In the introduction to this chapter, we spoke of comparing the ordinality of well-ordered sets. If A and B are well-ordered sets, we say that A has a *lower ordinality* than B ($A \prec B$) if A is isomorphic with an initial segment of B. It is convenient now to add: the ordinality of A is *less than or equal to* the ordinality of B if and only if A is isomorphic with B or an initial segment of B; in this case, we write $A \preccurlyeq B$. This is the same as saying that there exists an injective, order-preserving function from A to B, whose range is a section of B. (See 4.48 and 4.56.)

9.10 Lemma Let A and B be well-ordered sets. $A \preccurlyeq B$ if there exists an injective, order preserving function $f: A \to B$.

Proof. If $A \preccurlyeq B$ then clearly there exists an injective, order preserving function from A to B.

Conversely, suppose there exists an injective, order preserving function $f: A \to B$. If $C = \overline{f}(A)$, then $f: A \to C$ is an isomorphism. By 4.63, there exists an isomorphism $g: C \to D$, where D is B or an initial segment of B. Now $g \circ f: A \to D$ is an isomorphism, hence $A \preccurlyeq B$. ■

9.11 Corollary If $A \preccurlyeq B$ and $B \preccurlyeq C$ then $A \preccurlyeq C$.

Proof. Clearly the composite of two injective, order preserving functions is injective and order preserving. ■

If A and B are well-ordered sets and A has a lower ordinality than B, we quite naturally expect the ordinal number of A to be "less than" the ordinal number of B. Accordingly, we define the "natural" ordering of the ordinal numbers as follows.

9.12 Definition Let α and β be ordinals, and let A and B be well-ordered sets such that $\alpha = \circledcirc A$ and $\beta = \circledcirc B$. The relation $\leqslant$ is defined by

$$\alpha \leqslant \beta \qquad \text{if and only if} \qquad A \preccurlyeq B.$$

We note that $\alpha \leqslant \beta$ if and only if $\alpha \preccurlyeq \beta$.

The relation $\leqslant$ which we have just defined is obviously reflexive; it is antisymmetric by Lemma 4.61; it is transitive by Lemma 9.10; hence it is an order relation among the ordinal numbers. We can say more:

9.13 Theorem Every nonempty class of ordinal numbers has a least element.

Proof. Let $\mathcal{O}$ be a nonempty class of ordinal numbers, and let α be an arbitrary element of $\mathcal{O}$. If α is the least element of $\mathcal{O}$, we are done; otherwise, let $\mathscr{B} = \{\beta \in \mathcal{O} \mid \beta < \alpha\}$. It follows from our definition of the relation $<$ that every $\beta \in \mathscr{B}$ is similar to an initial segment of α. For each $\beta \in \mathscr{B}$, let $\phi(\beta)$ be the least element $x \in \alpha$ such that $\beta \approxeq S_x$. Now the set $\{\phi(\beta) \mid \beta \in \mathscr{B}\}$ has a least element $\phi(\delta)$ because it is a subset of α. We will show that δ is the least element of $\mathscr{B}$.

Indeed, let $\beta \in \mathscr{B}$; then $\phi(\delta) \leqslant \phi(\beta)$, hence $S_{\phi(\delta)} \subseteq S_{\phi(\beta)}$, so by 4.63 $S_{\phi(\delta)} \preccurlyeq S_{\phi(\beta)}$. Thus we have

$$\delta \approxeq S_{\phi(\delta)} \preccurlyeq S_{\phi(\beta)} \approxeq \beta,$$

so, by 9.10, $\delta \leqslant \beta$. ■

Thus, **the class of all the ordinal numbers is well ordered.**

9.14 Theorem

i) If $\beta > 0$, then $\alpha < \alpha + \beta$.
ii) $\beta \leqslant \alpha + \beta$.

Proof

i) Let A and B be well-ordered sets such that $\alpha = \circ A$ and $\beta = \circ B$. If b is the least element of B, then clearly A is the initial segment S_b of $A \oplus B$. Thus $A \prec A \oplus B$, so $\alpha < \alpha + \beta$.

ii) $B \subseteq A \oplus B$, hence by 4.63, B is isomorphic with $A \oplus B$ or an initial segment of $A \oplus B$. Thus $B \preccurlyeq A \oplus B$, so $\beta \leqslant \alpha + \beta$. ■

9.15 Theorem Let α and β be ordinals such that $\alpha < \beta$. Then there exists a unique ordinal $\gamma > 0$ such that $\alpha + \gamma = \beta$.

Proof. If A and B are well-ordered sets such that $\alpha = \circ A$ and $\beta = \circ B$, then $A \prec B$; that is, $A \approxeq S_x$ for some $x \in B$. Let $C = B - S_x$; C is well ordered and $C \neq \varnothing$, so if $\gamma = \circ C$, then $\gamma > 0$. Now $B = S_x \oplus C$, $\alpha = \circ S_x$ (because $S_x \approxeq A$), so $\beta = \alpha + \gamma$. For uniqueness, suppose $\beta = \alpha + \gamma = \alpha + \gamma'$, where, say, $\gamma < \gamma'$; that is, $\gamma' = \gamma + \delta$ $(\delta > 0)$. Then

$$\beta = \alpha + \gamma' = \alpha + \gamma + \delta = \beta + \delta \ (\delta > 0),$$

which is in contradiction with the result of Theorem 9.14(i). Thus $\gamma = \gamma'$. ■

9.16 Theorem For any ordinal numbers α, β, γ, the following rules hold:

i) $\alpha < \beta \Rightarrow \gamma + \alpha < \gamma + \beta$,
ii) $\gamma + \alpha < \gamma + \beta \Rightarrow \alpha < \beta$,
iii) $\alpha \leqslant \beta \Rightarrow \alpha + \gamma \leqslant \beta + \gamma$,
iv) $\alpha + \gamma < \beta + \gamma \Rightarrow \alpha < \beta$,
v) $\alpha < \beta, \gamma > 0 \Rightarrow \gamma\alpha < \gamma\beta$,
vi) $\gamma\alpha < \gamma\beta \Rightarrow \alpha < \beta$,
vii) $\alpha \leqslant \beta \Rightarrow \alpha\gamma \leqslant \beta\gamma$,
viii) $\alpha\gamma < \beta\gamma \Rightarrow \alpha < \beta$.

Proof. In (i), (iii), (v), and (vii) we assume that $\alpha < \beta$, hence we assume that there exists $\delta > 0$ such that $\beta = \alpha + \delta$. [*Note:* In (iii) and (vii), the case $\alpha = \beta$ is easily disposed of; indeed, if $\alpha = \beta$, then $\alpha + \gamma = \beta + \gamma$ and $\alpha\gamma = \beta\gamma$ (see Exercise 1, Exercise Set 9.2).]

i) $\gamma + \beta = \gamma + (\alpha + \delta) = (\gamma + \alpha) + \delta > \gamma + \alpha$ [this last relation is a consequence of 9.14(i)], so $\gamma + \beta > \gamma + \alpha$.

ii) Suppose $\gamma + \alpha < \gamma + \beta$. If $\alpha = \beta$, then $\gamma + \alpha = \gamma + \beta$ (Exercise 1, Exercise Set 9.2). If $\beta < \alpha$, then $\gamma + \beta < \gamma + \alpha$ by (i). Hence $\alpha < \beta$.

iii) Suppose, on the contrary, that $\beta + \gamma < \alpha + \gamma$; that is, $\alpha + (\delta + \gamma) < \alpha + \gamma$. Then $\delta + \gamma < \gamma$ by (ii), and this is impossible by 9.14(ii); thus $\alpha + \gamma \leqslant \beta + \gamma$.

iv) Suppose $\alpha + \gamma < \beta + \gamma$. If $\alpha = \beta$, then $\alpha + \gamma = \beta + \gamma$. If $\beta < \alpha$, then $\beta + \gamma \leqslant \alpha + \gamma$ by (iii). Thus $\alpha < \beta$.

v) $\gamma\beta = \gamma(\alpha + \delta) = \gamma\alpha + \gamma\delta > \gamma\alpha$ [this last relation holds by 9.14(i)].

vi) Suppose $\gamma\alpha < \gamma\beta$. If $\alpha = \beta$, then $\gamma\alpha = \gamma\beta$. If $\beta < \alpha$, then $\gamma\beta < \gamma\alpha$ by (v). Thus $\alpha < \beta$.

vii) We must show that $\alpha\gamma \leqslant (\alpha + \delta)\gamma$. Let $\alpha = \circledcirc A$, $\gamma = \circledcirc C$, $\delta = \circledcirc D$; $A \subseteq A \oplus D$, so

$$A \odot C \subseteq (A \oplus D) \odot C.$$

It follows by 4.63 that

$$A \odot C \prec\!\!\!\leqslant (A \oplus D) \odot C,$$

or $\alpha\gamma \leqslant (\alpha + \delta)\gamma$.

viii) Suppose $\alpha\gamma < \beta\gamma$. If $\alpha = \beta$, then $\alpha\gamma = \beta\gamma$. If $\beta < \alpha$, then $\beta\gamma \leqslant \alpha\gamma$ by (vii). Hence $\alpha < \beta$. ■

9.17 Theorem

i) If $\gamma + \alpha = \gamma + \beta$, then $\alpha = \beta$.

ii) Assume $\gamma > 0$. If $\gamma\alpha = \gamma\beta$, then $\alpha = \beta$.

The proof, an immediate consequence of 9.16(i) and (v), is left as an exercise for the reader.

9.18 Lemma If $\gamma < \beta\alpha$, then there exist ordinals δ and ε such that $\gamma = \beta\delta + \varepsilon$, $\delta < \alpha$, and $\varepsilon < \beta$.

Proof. Let A, B, and C be well-ordered sets such that $\alpha = \circledcirc A$, $\beta = \circledcirc B$ and $\gamma = \circledcirc C$. Our assumption is that $C \prec B \odot A$, that is, $C \cong S_{(b,a)}$ for some $(b, a) \in B \odot A$. Let $E = \{(x, a) \mid x < b\}$, that is, $E = S_b \odot \{a\}$; clearly $E \cong S_b$. We will show that

$$S_{(b,a)} = (B \odot S_a) \oplus E$$

(this relation is illustrated in Fig. 9.2).

Let $x \in B$, $y \in A$. Then

$$\begin{aligned}(x, y) \in S_{(b,a)} &\Leftrightarrow (x, y) < (b, a)\\ &\Leftrightarrow y < a \text{ (that is, } y \in S_a) \text{ or } [y = a \text{ and } x < b]\\ &\Leftrightarrow (x, y) \in B \times S_a \text{ or } (x, y) \in E\\ &\Leftrightarrow (x, y) \in (B \times S_a) \cup E\end{aligned}$$

Thus $S_{(b,a)} = (B \times S_a) \cup E$.

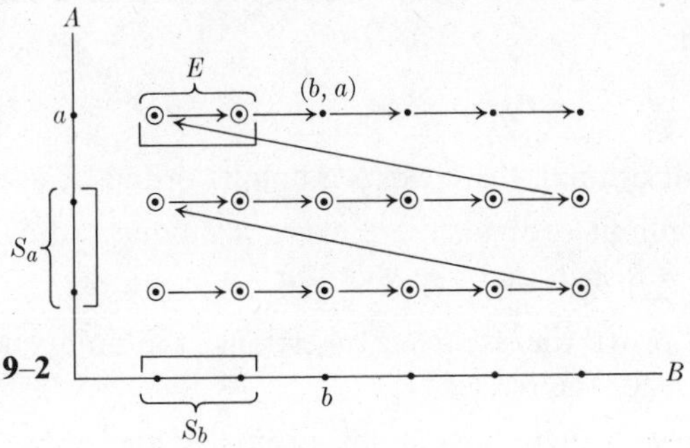

Fig. 9–2

$\odot\odot\odot\odot$ denotes $S_{(b,a)}$

Now it is easy to verify that the ordering of $(B \odot S_a) \oplus E$ is the same as the ordering of $S_{(b,a)}$; the details are left as an exercise for the reader. We conclude that

$$S_{(b,a)} = (B \odot S_a) \oplus E.$$

Let $\varepsilon = \circledcirc E = \circledcirc S_b$ and $\delta = \circledcirc S_a$. By the definition of the relation $<$, $\varepsilon < \beta$ and $\delta < \alpha$. Now $\gamma = \circledcirc C = \circledcirc S_{(b,a)}$; thus $\gamma = \beta\delta + \varepsilon$. ■

It is very useful to note that the "division algorithm" for the natural numbers can be generalized to all the ordinal numbers.

9.19 Theorem If α and $\beta > 0$ are ordinals, then there exist unique ordinals ξ and ρ such that $\alpha = \beta\xi + \rho$ and $\rho < \beta$.

Proof

Existence. Since $\beta \geqslant 1$, we have $\beta\alpha \geqslant \alpha$. If $\beta\alpha = \alpha$, we are done; otherwise, $\alpha < \beta\alpha$, so by Lemma 9.18 there exist ordinals $\delta < \alpha$ and $\varepsilon < \beta$ such that $\alpha = \beta\delta + \varepsilon$, hence again we are done.

Uniqueness. Assume $\alpha = \beta\xi + \rho = \beta\xi' + \rho'$, where (say) $\xi' < \xi$; that is, $\xi = \xi' + \mu$ $(\mu > 0)$. Then

$$\beta\xi' + \rho' = \beta\xi + \rho = \beta(\xi' + \mu) + \rho = \beta\xi' + \beta\mu + \rho,$$

so $\rho' = \beta\mu + \rho > \beta\mu \geqslant \beta$, which contradicts our assumption that $\rho' < \beta$. Thus we cannot have $\xi' < \xi$; by symmetry, we cannot have $\xi < \xi'$; thus $\xi = \xi'$. It follows by 9.17(i) that $\rho = \rho'$. ■

If α is an ordinal number, it is easy to see that $\alpha + 1$ is the immediate successor of α. Now let β be a non-zero ordinal number; if β has no immediate predecessor—that is, if β is not equal to $\alpha + 1$ for any ordinal α—then β is called a *limit ordinal.* Otherwise—that is, if β has an immediate predecessor—then β is called a *nonlimit ordinal.* Limit ordinals have the following useful properties.

9.20 Theorem

i) If α is a limit ordinal, there exists a unique ordinal ξ such that $\alpha = \omega\xi$.

ii) If α is a nonlimit ordinal, there exists a unique ordinal ξ and a unique finite ordinal $n \neq 0$ such that $\alpha = \omega\xi + n$.

Proof. We will prove the existence assertions; the uniqueness assertions are left as an exercise for the reader.

i) By Theorem 9.19, there exist unique ordinals ξ and ρ such that $\alpha = \omega\xi + \rho$ and $\rho < \omega$. But if $\rho < \omega$, then ρ must be finite. But then ρ must be 0; for otherwise $\rho = m + 1$ for some finite m, hence $\alpha = \omega\xi + m + 1$ would not be a limit ordinal.

ii) As above, $\alpha = \omega\xi + \rho$, where ρ is finite. Now $\rho \neq 0$; for if $\rho = 0$, then $\alpha = \omega\xi$, which is impossible because $\omega\xi$ is a limit ordinal (see Exercise 6, Exercise Set 9.3). ■

EXERCISES 9.3

1. Prove that $1 + \alpha = \alpha$ if and only if $\alpha \geqslant \omega$.
2. An ordinal number $\rho > 0$ is called *irreducible* if there exists no pair of ordinals α, β such that $\alpha < \rho$, $\beta < \rho$, and $\alpha + \beta = \rho$. Prove the following:
   a) An ordinal ρ is irreducible if and only if $\pi + \rho = \rho$ for every ordinal $\pi < \rho$.
   b) Suppose $\rho > 1$ and $\varepsilon > 0$; $\varepsilon\rho$ irreducible $\Rightarrow \rho$ irreducible.
   c) If ρ is irreducible and $0 < \mu < \rho$, then there exists an irreducible ordinal ξ such that $\rho = \mu\xi$.
   d) Suppose $\alpha > 0$; the set of all irreducible ordinals $\leqslant\alpha$ has a greatest element. [*Hint:* Consider the set of all β such that $\alpha = \rho + \beta$ for some irreducible ρ.]
3. Show that an ordinal α is a limit ordinal if and only if

$$\beta < \alpha \Rightarrow (\beta + 1) < \alpha.$$

4. Let γ be a *non*limit ordinal. Prove the following.

a) $\alpha < \beta \Rightarrow \alpha\gamma < \beta\gamma$, b) $\alpha\gamma = \beta\gamma \Rightarrow \alpha = \beta$.

5. Let $\beta \neq 0$. Prove that $\alpha + \beta$ is a limit ordinal if and only if β is a limit ordinal.

6. Let $\alpha, \beta \neq 0$. Prove that $\alpha\beta$ is a limit ordinal if and only if α is a limit ordinal or β is a limit ordinal. [*Hint:* Use Exercise 3 and Lemma 9.18.]

7. Prove that $n\omega = \omega$, $\forall n \in \omega$. [*Hint:* Use Theorem 9.19 to "divide" ω by n.]

8. Let $\alpha \neq 0$. Prove that α is a limit ordinal if and only if $n\alpha = \alpha$, for every finite n. [*Hint:* Use Exercise 7 and Theorem 9.20(i). For the converse, use Exercise 4(b).]

9. a) Use induction to prove that if γ is an infinite ordinal, then $(\gamma + 1)n = \gamma n + 1$ for all finite n. [Use Exercise 1.]

   b) Prove that $\forall \gamma > 0$, $(\gamma + 1)\omega = \gamma\omega$. [*Hint:* If γ is infinite, assume $\gamma\omega < (\gamma + 1)\omega$ and use Theorem 9.18 to arrive at a contradiction. If γ is finite, use Exercise 7.]

   c) Conclude that if β is any limit ordinal, then
   $$(\gamma + 1)\beta = \gamma\beta, \quad \forall \gamma > 0.$$

10. Let α be a *non*limit ordinal. Prove that $\forall \gamma > 0$, $(\gamma + 1)\alpha > \gamma\alpha$. [*Hint:* Use Exercise 4.]

11. If α is an infinite ordinal and $\beta \neq 0$ is a *non*limit ordinal, prove that
$$(\alpha + 1)\beta = \alpha\beta + 1.$$
[*Hint:* Use Theorem 9.20(ii) and assume $\alpha\beta + 1 < (\alpha + 1)\beta$ to arrive at a contradiction. Use Exercise 9.]

12. a) If α is a limit ordinal, prove that $\alpha = \sup\{\beta \mid \beta < \alpha\}$. [Use Exercise 3.]

    b) If α is a limit ordinal and β is any ordinal, prove that
    $$\beta + \alpha = \sup\{\beta + \mu \mid \mu < \alpha\}.$$
    [*Hint:* If γ is an upper bound of $\{\beta + \mu \mid \mu < \alpha\}$, then $\gamma > \beta$, that is, $\gamma = \beta + \delta$; δ proves to be an upper bound of $\{\mu \mid \mu < \alpha\}$.]

13. If α is a limit ordinal and β is any ordinal, prove that
$$\beta\alpha = \sup\{\beta\mu \mid \mu < \alpha\}.$$
[*Hint:* If γ is an upper bound of $\{\beta\mu \mid \mu < \alpha\}$, then $\gamma = \alpha\delta + \rho$ $(\rho < \alpha)$. Note that $\mu < \alpha \Rightarrow \gamma \geqslant \alpha(\mu + 1)$ and conclude that δ is an upper bound of $\{\mu \mid \mu < \alpha\}$.]

14. Prove Theorem 9.17.

15. Prove the uniqueness assertions of Theorem 9.20.

4 THE ALEPHS AND THE CONTINUUM HYPOTHESIS

In Chapter 8 it was proven that the relation $\leqslant$ among cardinals is a well-ordering; hence there is a smallest infinite cardinal, a next greater infinite cardinal, and so on; every infinite cardinal has a uniquely determined immediate successor. It follows that the infinite cardinal numbers can be ranked in "first, second, third, ..." order. This opportunity of ranking the cardinals—and using the ordinals to designate the ranks—has valuable applications in mathematics.

We will proceed to show that there is an isomorphism between the class of all the infinite cardinals and the class of all the ordinals.

9.21 Theorem Let IC be the class of all the infinite cardinals and let OR be the class of all the ordinals. There exists an isomorphism

$$\Omega\colon \mathrm{IC} \to \mathrm{OR}$$

Proof. We begin by proving the following:

i) Every initial segment of IC is a set.

Let a be an infinite cardinal and let $B = \{b \mid b \in \mathrm{IC} \text{ and } b \leqslant a\}$. We well-order a, and then define a function $\phi\colon a \to B$ by

$$\phi(x) = \# S_x, \quad \forall x \in a.$$

It is easy to see that ϕ is surjective, for if $c \in B$, that is, c is an infinite cardinal and $c \leqslant a$, then c is equipotent with a subset $C \subseteq a$. Thus by 4.63, c is equipotent with an initial segment $S_x \subseteq a$, so $c = \# S_x = \phi(x)$. Now a is a set, hence by Axiom A7, B is a set.

Using an argument analogous to the above, we can prove

ii) Every initial segment of OR is a set.

Now IC and OR are both well-ordered classes, hence by 4.62, exactly one of the following three cases must hold: (a) IC $\approxeq$ OR, (b) IC is similar to an initial segment of OR, (c) OR is similar to an initial segment of IC. Suppose for a moment that (c) holds. By (ii), every initial segment of IC is a set, hence by 2.36, OR is a set. But we have proved (page 168) that OR is a proper class, hence (c) cannot hold. Analogously, (b) cannot hold, which proves that (a) holds. ■

If a is an infinite cardinal number, the ordinal $\Omega(a)$ is called the *ordinal rank* of a. Note that Ω is an isomorphism; thus, if a is the least infinite cardinal, then $\Omega(a) = 0$; if a is the next greater infinite cardinal, then $\Omega(a) = 1$, and so on.

The infinite cardinals are often called *alephs*. If a is an infinite cardinal and $\alpha = \Omega(a)$, we frequently write

$$a = \aleph_\alpha.$$

Thus the first few infinite cardinals are $\aleph_0, \aleph_1, \aleph_2, \dots$

Theorem 9.21 has the following simple consequences:

9.22 i) $\aleph_\alpha = \aleph_\beta \Rightarrow \alpha = \beta,$

ii) $\alpha = \beta \Rightarrow \aleph_\alpha = \aleph_\beta$,
iii) $\aleph_\alpha < \aleph_\beta \Rightarrow \alpha < \beta$,
iv) $\alpha < \beta \Rightarrow \aleph_\alpha < \aleph_\beta$.

We have seen that every infinite cardinal number a has an immediate successor—but what exactly *is* the immediate successor of a? We know that 2^a is greater than a, but is there any cardinal number between a and 2^a? Let us ask a more specific question: We have seen that $\aleph_0$ is the cardinal number of denumerable sets (for the cardinal number of ω is the least infinite cardinal), and that $2^{\aleph_0}$ is the cardinal number of the real numbers; is there a cardinal number between these two?

The early set theorists proposed the hypothesis that there is no cardinal between $\aleph_0$ and $2^{\aleph_0}$, and named it the *continuum hypothesis*—for it is equivalent to saying that every set of real numbers which is not denumerable has the power of the real numbers, called the "power of the continuum." An obvious extension of this conjecture is the statement: For every infinite cardinal number a, there is no cardinal between a and 2^a; this is called the *generalized continuum hypothesis.*

Continuum Hypothesis. There does not exist any cardinal c such that $\aleph_0 < c < 2^{\aleph_0}$.

Generalized Continuum Hypothesis. If a is any infinite cardinal, there does not exist any cardinal c such that $a < c < 2^a$.

It has been proven in recent years that the continuum hypothesis and the generalized continuum hypothesis cannot be proven from the other axioms of set theory, and do not contradict these. Hence their status is analogous to that of Euclid's "Fifth Postulate" in geometry. We may postulate them or deny them, in each case getting a consistent theory of cardinal numbers.

EXERCISES 9.4

1. Prove that the generalized continuum hypothesis is equivalent to

$$2^{\aleph_\alpha} = \aleph_{\alpha+1}.$$

2. Assuming the generalized continuum hypothesis, prove the following:

$$\text{If } \alpha \leqslant \beta, \quad \text{then} \quad \aleph_\alpha^{\aleph_\beta} = \aleph_{\beta+1}.$$

3. Assuming the generalized continuum hypothesis, prove the following:

For arbitrary cardinals a, b, $\quad a < b \Rightarrow 2^a < 2^b$.

Further problems on the alephs are given in Exercise Set 9.5.

*5 CONSTRUCTION OF THE ORDINALS AND CARDINALS

We said, in the introduction to this chapter, that it is possible to construct sets which satisfy Conditions 01 and 02 of the Axiom of Ordinality. The chief purpose of this construction is to prove that we can dispense with the Axiom of Ordinality by actually producing the sets whose existence the axiom asserts.

Our process of construction is based upon the same idea—outlined on page 125—that we used to construct the natural numbers. We begin by defining

$$0 = \varnothing,$$
$$1 = \{\varnothing\},$$
$$2 = \{\varnothing, \{\varnothing\}\}, \quad \text{etc.}$$

If A is a set, we define the *successor of* A to be the set A^+, given by

$$A^+ = A \cup \{A\}.$$

Thus $0 = \varnothing$, $1 = 0^+$, $2 = 1^+$, and so on. This time, however, we will go further than we did in Chapter 7. Starting with ω, we define

$$\omega + 1 = \omega^+,$$
$$\omega + 2 = (\omega + 1)^+, \quad \text{etc.}$$

Then, starting with $\omega + \omega = \omega 2$, we get

$$\omega 2 + 1, \omega 2 + 2, \ldots, \omega 3, \omega 3 + 1, \ldots, \omega 4, \ldots, \ldots, \omega\omega,$$

and so on.

This is the *basic idea* of our construction process, but we will not proceed exactly in this fashion. Instead of starting with 0 and constructing successive sets one by one, we will define all the sets simultaneously. This can be accomplished in the following way.

The "elementhood" relation $\in$ is not, generally speaking, an order relation; for example, if $x \in A$ and $A \in B$, it does not necessarily follow that $x \in B$. However, there are special cases where $\in$ does behave as if it were an order relation; one of these cases concerns us here.

It will be convenient to consider, rather than $\in$, the relation $\underline{\in}$ defined as follows:

9.23 $x \underline{\in} y$ if and only if $x = y$ or $x \in y$.

9.24 Definition A class A is said to be $\in$-*ordered* if it is ordered by the relation $\underline{\in}$ and if $\forall x \in A, x \notin x$.

A class A is said to be $\in$-*well-ordered* if it is well-ordered by the relation $\underline{\in}$ and if $\forall x \in A, x \notin x$.

* This section may be omitted without loss of continuity.

If A is $\in$-ordered, then clearly "x is less than or equal to y" means that $x \in y$ or $x = y$; "x is strictly less than y" means that $x \in y$. It is clear that any subclass of an $\in$-well-ordered class is $\in$-well-ordered.

As we stated earlier (Definition 6.5), a set A is said to be *transitive* if $x \in A \Rightarrow x \subseteq A$.

9.25 Definition A transitive, $\in$-well-ordered set will (temporarily) be called an *ord*.

Examples of ords are the natural numbers. They are transitive sets by Lemma 6.6; it follows immediately from Definition 6.21 and Theorem 6.25 that every natural number is $\in$-well-ordered (note simply that every natural number is a subset of ω). But ω, too, is an ord, and so are the sets $\omega + 1, \ldots, \omega 2, \ldots, \omega 3, \ldots$, and so on.

The name "ord" is a provisional one. The purpose of this section is to prove that the ords satisfy 01 and 02; once this has been shown, they will be renamed "ordinal numbers."

9.26 Lemma Let α be an ord:

i) If $x \in \alpha$, then $x =$ the initial segment S_x of α.
ii) If $x \in \alpha$, then x is an ord.

Proof

i) If $u \in x$, then (because α is a transitive set) $u \in \alpha$. Since $\in$ is the order relation in α, $u \in x$ means that u is "less than" x, so $u \in S_x$. Conversely, if $u \in S_x$, then u is "less than" x, that is, $u \in x$; thus $x = S_x$.
ii) Let $x \in \alpha$. By (i), $x = S_x$, hence (because $S_x \subseteq \alpha$) x is $\in$-well-ordered. It remains to show that x is a transitive set. Suppose $v \in x$ and $u \in v$; now $v \in x$, $x \in \alpha$ and α is a transitive set, so $v \in \alpha$; analogously, $u \in \alpha$. Thus u, v, and x are elements of α, with $u \in v$ and $v \in x$; but α is ordered by $\in$, hence $u \in x$. This proves that $v \subseteq x$; thus x is a transitive set. ■

9.27 Theorem Let α and β be ords. If $\alpha \approxeq \beta$ then $\alpha = \beta$.

Proof. By hypothesis, there exists an isomorphism $f : \alpha \to \beta$. Let

$$A = \{x \in \alpha \mid f(x) = x\}.$$

We will prove that $A = \alpha$, whence it follows immediately that $\alpha = \beta$. If $A \neq \alpha$, let c be the least element of $\alpha - A$. Thus ■

9.28 $\forall x \in c, f(x) = x$, whereas $f(c) \neq c$.

Now,

$$x \in c \Rightarrow f(x) \in f(c) \Rightarrow x \in f(c) \qquad [\text{because } x = f(x)].$$

Conversely,

$$\begin{aligned} x \in f(c) &\Rightarrow x \in \beta \qquad [\text{because } f(c) \in \beta \text{ and } \beta \text{ is a transitive set}] \\ &\Rightarrow x = f(y) \qquad \text{for some } y \in \alpha \\ &\Rightarrow f(y) \in f(c) \qquad [\text{because } x \in f(c)] \\ &\Rightarrow y \in c \qquad [\text{because } f \text{ is an isomorphism}] \\ &\Rightarrow f(y) = y \qquad [\text{by 9.28}]. \end{aligned}$$

Thus, briefly, $x = f(y) = y \in c$. Hence $x \in c$ iff $x \in f(c)$, that is, $f(c) = c$, contrary to our choice of c. This contradiction proves that $A = \alpha$. ■

Let $\langle \text{ORD}, \underline{\in} \rangle$, or briefly ORD, designate the class of all the ords with the relation $\underline{\in}$.

9.29 Theorem ORD is an $\in$-well-ordered class.

Proof. We must prove that $\underline{\in}$ is a well-ordering of the class ORD.

i) For every ord α, $\alpha = \alpha$, hence $\alpha \underline{\in} \alpha$; thus $\underline{\in}$ is reflexive.

ii) Suppose $\alpha \underline{\in} \beta$ and $\beta \underline{\in} \alpha$; if $\alpha = \beta$, we are done; we will show that $\alpha \neq \beta$ is impossible. Indeed, if $\alpha \neq \beta$, then $\alpha \in \beta$ and $\beta \in \alpha$, hence $\alpha \subseteq \beta$ and $\beta \subseteq \alpha$, so $\alpha = \beta$; this contradicts the assumption that $\alpha \neq \beta$.

iii) Suppose that $\alpha \underline{\in} \beta$ and $\beta \underline{\in} \gamma$. It follows immediately, from the fact that γ is a transitive set, that $\alpha \underline{\in} \gamma$; thus $\underline{\in}$ is transitive.

iv) Let α be an ord. If $\alpha \in \alpha$, then α is an element of the $\in$-ordered set α, hence by 9.24, $\alpha \notin \alpha$. Thus, for every $\alpha \in \text{ORD}$, $\alpha \notin \alpha$.

v) We will show next that any two ords α, β are comparable; by 4.62, $\alpha \approxeq \beta$, or α is similar to an initial segment of β, or β is similar to an initial segment of α. By 9.26 and 9.27, it follows immediately that α and β are comparable.

vi) We will now show that every nonempty subclass of ORD has a least element. Let A be a nonempty subclass of ORD, and let $\alpha \in A$. If α is the least element of A, we are done; otherwise, assume α is not the least element of A. Then there exists an element $\delta \in A$ such that δ is "less than" α, that is, $\delta \in \alpha$. Thus $A \cap \alpha \neq \varnothing$. Since $A \cap \alpha$ is a nonempty subclass of α, it follows that $A \cap \alpha$ has a least element γ. We will prove that γ is the least element of A. For otherwise [note that by (v), γ is comparable with all the elements of A], there is a $\beta \in A$ such that β is "less than" γ, that is, $\beta \in \gamma$. Now $\beta \in \gamma$ and $\gamma \in \alpha$, hence $\beta \in \alpha$ because α is a transitive set, so $\beta \in A \cap \alpha$. But this is impossible because γ was chosen to be the least element of $A \cap \alpha$. This proves that γ is the least element of A. ■

9.30 Lemma If $\alpha \in$ ORD, then $\alpha =$ the initial segment S_α of ORD.

The proof of this lemma is trivial, and is left as an exercise for the reader.

9.31 Theorem ORD is a proper class.

Proof. Suppose, on the contrary, that ORD is a set. Then by 9.29 and 9.26(ii), ORD is an ord, hence ORD $\in$ ORD. But this is impossible, because by the definition of $\in$-ordered, if x is an element of an ord, then $x \notin x$. ■

9.32 Theorem Let A be an arbitrary well-ordered set; then there exists an ord α such that $A \approxeq \alpha$.

Proof. If A is a well-ordered set, then by 4.62,

i) $A \approxeq$ ORD, or

ii) ORD is similar to an initial segment of A, or

iii) A is similar to an initial segment of ORD.

Cases (i) and (ii) are impossible by 9.31, hence case (iii) holds. By 9.30, this means that A is similar to an element $\alpha \in$ ORD. ■

Theorems 9.27 and 9.32 assert that the class ORD satisfies Conditions 01 and 02. We are justified, therefore, in renaming the ords *ordinal numbers*, and writing ORD = OR.

Since OR = ORD, every ordinary number is an "ord" as defined in 9.25. In particular, since every natural number is an "ord" and ω itself is an "ord," it follows that every natural number is an ordinal number and ω is an ordinal number. Now it is easy to prove that if X is an "ord" then $X^+ = X \cup \{X\}$ is an "ord," that is, if α is an ordinal number then α^+ is an ordinal number. (Note that α^+ is obtained by adjoining a single element to α, hence $\alpha^+ = \alpha + 1$.) Thus, the class OR begins with the following elements:

$$0, 1, 2, \ldots, \omega, \omega + 1, \omega + 2, \ldots, \omega + \omega, \ldots$$

If A is an arbitrary set, consider the class of all the ordinals equipotent with A; this class has a unique least element, which we call the *initial ordinal* equipotent with A. It is trivial, now, to verify that the class of all the initial ordinals satisfies Conditions K1 and K2 of the Axiom of Cardinality. Thus we are justified in making the following definition.

9.33 Definition By a *cardinal number* we mean an initial ordinal.

Thus, the class CD of the cardinal numbers is the class of all the initial ordinals.

We have thus fulfilled our promise of actually producing sets to serve as the cardinal numbers and the ordinal numbers.

We noted earlier that every natural number is a transitive, $\in$-well-ordered set, that is, an ordinal number. It is immediate, too, that every natural number is an *initial* ordinal number. Thus, in our construction, the natural numbers coincide with the finite ordinals, as well as with the finite cardinals.

The reader should note that everything we have proved in Chapters 8 and 9 about the class CD of the cardinal numbers has depended solely upon Conditions K1 and K2 of the Axiom of Cardinality. Thus everything we have already said about the cardinals holds, without any alteration, for the class CD defined by 9.33. In particular, 9.22 still holds, that is,

$$\alpha \leftrightarrow \aleph_\alpha$$

is an isomorphism between OR and the class of all the infinite cardinals.

It is important to note that by 9.33, $\aleph_\alpha$ is both a cardinal and an ordinal (specifically, an initial ordinal). In order to avoid any confusion, it is common practice in mathematics to write

$$\aleph_\alpha = \omega_\alpha$$

for every infinite cardinal $\aleph_\alpha$, and to treat $\aleph_\alpha$ as a cardinal number and ω_α as an ordinal number. In other words, the number in question is denoted by $\aleph_\alpha$ when it is used as a cardinal, and by ω_α when it is used as an ordinal. For example, $\aleph_\alpha + \aleph_\beta$ designates the cardinal sum of the two numbers, whereas $\omega_\alpha + \omega_\rho$ designates their ordinal sum.

EXERCISES 9.5

1. For each ordinal number α, prove that $\alpha \leqslant \omega_\alpha$; conclude that $\#\alpha \leqslant \aleph_\alpha$. [*Hint:* Use 4.58.]

2. Prove that if β is a limit ordinal, then $\aleph_\beta = \sup\{\aleph_\gamma \mid \gamma < \beta\}$. [*Hint:* Use 9.22.]

3. Prove that $\sum_{\gamma \leqslant \mu} \aleph_\gamma = \aleph_\mu$.

4. If μ is a limit ordinal, prove that $\sum_{\gamma < \mu} \aleph_\gamma = \aleph_\mu$.

10

*Transfinite Recursion. Selected Topics in the Theory of Ordinals and Cardinals

1 TRANSFINITE RECURSION

In Chapter 6 we discussed the notions of finite induction and finite recursion; finite induction is a method of *proof by induction* in ω, and finite recursion is a method of *definition by induction* in ω. Let us briefly review these two concepts. Proof by induction in ω works as follows: A theorem is first shown to be true for 0. Then it is shown that if the theorem holds for n, it must hold for $n + 1$. We are able to conclude that the theorem holds for all the natural numbers. In order to define a function by induction we proceed in much the same manner. First, we define the value of the function at 0. Then we use its value at n to define its value at $n + 1$. In this way its value is defined at every natural number.

We have already noted that OR, the class of the ordinal numbers, is an extension of ω. The finite ordinal numbers are, respectively, 0, 1, 2, 3, ...; after these comes ω, then $\omega + 1$, $\omega + 2$, ..., $\omega 2$, $\omega 2 + 1$, ..., and so on. Some ordinal numbers (other than 0) have no immediate predecessor; we call them *limit ordinals.* Because of them, finite induction cannot be used to prove theorems for all the elements of OR. Indeed, suppose we are able to show that a theorem holds for 0, and that whenever it holds for α it holds also for $\alpha + 1$; we may then conclude that the theorem holds for all the finite ordinals 0, 1, 2, ..., but not necessarily for ω (because finite recursion permits us to make an inference from any number α to its immediate successor $\alpha + 1$, but ω is not the immediate successor of any number). Furthermore, we may not conclude that the theorem necessarily holds for any ordinal greater than ω. Proof by induction in OR must use the principle of *transfinite induction*: We show that if a theorem holds for all the elements less than any given β, then it must hold also for β; we may then conclude (see Theorem 4.57) that it holds for all the ordinals.

The situation is analogous when we wish to use transfinite induction to define a function f on OR. We may begin by defining the value of f at 0. Next, we may use its value at α to define its value at $\alpha + 1$; this will define the values of f at 0, 1, 2, ... But in order to "reach beyond" the finite ordinals we need an additional rule of definition, such as the following: If β is any

* This chapter may be omitted without loss of continuity.

limit ordinal, then

$$f(\beta) = \sup\{f(\alpha) \mid \alpha < \beta\}.$$

With this additional rule it would appear to us that all the values of f (at all the ordinal numbers) are determined. This is indeed true, and the statement of this fact is called the *transfinite recursion theorem*; it will be proven next.

(In order to clearly understand the meaning of the transfinite recursion theorem we advise the reader to compare it with Theorem 6.8—the finite recursion theorem—and reread the comments which precede 6.8.)

10.1 Transfinite Recursion Theorem Let A be a well-ordered class, let $f: A \to A$ be a function, and let $a \in A$. Then there exists a unique function u: OR $\to A$ such that

i) $u(0) = a$,
ii) $u(\alpha^+) = f(u(\alpha))$, $\forall \alpha \in \text{OR}$,*
iii) $u(\alpha) = \sup\{u(\gamma) \mid \gamma < \alpha\}$ if α is a limit ordinal.

Proof. The proof is an adaptation of the argument we used to prove 6.8. If $G \subseteq \text{OR} \times A$, we will call G *perfect* if it satisfies the following conditions:

i′) $(0, a) \in G$.
ii′) $\forall \alpha \in \text{OR}$, if $(\alpha, x) \in G$, then $(\alpha^+, f(x)) \in G$.
iii′) For every limit ordinal α, if $(\gamma, x_\gamma) \in G$ for all $\gamma < \alpha$ and if $x = \sup\{x_\gamma \mid \gamma < \alpha\}$, then $(\alpha, x) \in G$.

It is clear that (i′), (ii′), (iii′) are not different from (i), (ii), (iii). There *are* perfect subclasses of OR $\times$ A, for example OR $\times$ A itself. Now, let u be the intersection of all the perfect subclasses of OR $\times$ A; that is,

$$u = \{(\alpha, x) \mid G \text{ is perfect} \Rightarrow (\alpha, x) \in G\}.$$

It is trivial to show that u is perfect; we will now prove that u is a function.

Let $S = \{\gamma \in \text{OR} \mid (\gamma, x) \in u \text{ for exactly one } x \in A\}$. Using induction, we will prove that $S = \text{OR}$, and it will follow that u is a function from OR to A. Suppose that $\forall \gamma < \beta$, $\gamma \in S$; that is, $\forall \gamma < \beta$, $(\gamma, x_\gamma) \in u$ for exactly one element $x_\gamma \in A$. We will consider two cases:

a) β is a nonlimit ordinal, $\beta = \delta^+$. Then $(\delta, x_\delta) \in u$, hence by (ii′),

$$(\delta^+, f(x_\delta)) = (\beta, f(x_\delta)) \in u.$$

Suppose now that $(\beta, y) \in u$ where $y \neq f(x_\delta)$; we will show that $u' = u - \{(\beta, y)\}$ is perfect, which is impossible because of the way we defined u, hence $(\beta, y) \notin u$.

u′ satisfies (i′): Note that u' is obtained from u by deleting only the one ordered pair (β, y). Now $\beta \neq 0$, so $(\beta, y) \neq (0, a)$; thus $(0, a) \in u'$.

---

* Note that α^+ is $\alpha + 1$.

u′ satisfies (ii′): Suppose $(\alpha, x) \in u'$; then $(\alpha, x) \in u$, so $(\alpha^+, f(x)) \in u$. We will consider two cases, according as $\alpha^+ \neq \beta$ or $\alpha^+ = \beta$. If $\alpha^+ \neq \beta$, then

$$(\alpha^+, f(x)) \neq (\beta, y),$$

so $(\alpha^+, f(x)) \in u'$. If $\alpha^+ = \beta = \delta^+$, then $\alpha = \delta$, so $(\alpha, x) = (\delta, x_\delta)$, and therefore

$$(\alpha^+, f(x)) = (\delta^+, f(x_\delta)).$$

But we are assuming that $y \neq f(x_\delta)$, hence $(\beta, y) \neq (\alpha^+, f(x))$, so $(\alpha^+, f(x)) \in u'$.

u′ satisfies (iii′): If α is a limit ordinal and $(\gamma, x_\gamma) \in u'$ for all $\gamma < \alpha$, then $(\gamma, x_\gamma) \in u$ for all $\gamma < \alpha$, so by (iii′),

$$(\alpha, \sup\{x_\gamma\}) \in u.$$

But $(\alpha, \sup\{x_\gamma\}) \neq (\beta, y)$ because β is a nonlimit ordinal, hence

$$(\alpha, \sup\{x_\gamma\}) \in u'.$$

b) β is a limit ordinal; then $(\gamma, x_\gamma) \in u$ for all $\gamma < \beta$, hence by (iii′), $(\beta, \sup\{x_\gamma \mid \gamma < \beta\}) \in u$. Suppose now that $(\beta, y) \in u$ where $y \neq \sup\{x_\gamma\}$. We will show that $u' = u - \{(\beta, y)\}$ is perfect, which is impossible because of the way we defined u, hence $(\beta, y) \notin u$.

u′ satisfies (i′): Certainly $\beta \neq 0$, hence $(\beta, y) \neq (0, a)$, so $(0, a) \in u'$.

u′ satisfies (ii′): If $(\alpha, x) \in u'$, then $(\alpha, x) \in u$, so by (ii′), $(\alpha^+, f(x)) \in u$. But $\alpha^+ \neq \beta$ because β is a limit ordinal, hence

$$(\alpha^+, f(x)) \neq (\beta, y),$$

so $(\alpha^+, f(x)) \in u'$.

u′ satisfies (iii′): If α is a limit ordinal and $(\gamma, x_\gamma) \in u'$ for all $\gamma < \alpha$, then $(\gamma, x_\gamma) \in u$ for all $\gamma < \alpha$, so

$$(\alpha, \sup\{x_\gamma \mid \gamma < \alpha\}) \in u.$$

If $\alpha \neq \beta$, then

$$(\alpha, \sup\{x_\gamma \mid \gamma < \alpha\}) \neq (\beta, y),$$

so $(\alpha, \sup\{x_\gamma \mid \gamma < \alpha\}) \in u'$. If $\alpha = \beta$, then

$$(\alpha, \sup\{x_\gamma \mid \gamma < \alpha\}) = (\beta, \sup\{x_\gamma \mid \gamma < \beta\}).$$

But we have assumed that $y \neq \sup\{x_\gamma \mid \gamma < \beta\}$, so

$$(\alpha, \sup\{x_\gamma \mid \gamma < \alpha\}) \neq (\beta, y),$$

hence $(\alpha, \sup\{x_\gamma\}) \in u'$.

We have proven that u is a function which satisfies conditions (i), (ii), and (iii). The proof of the uniqueness of u is left as an exercise for the reader. ■

Just as finite recursion was used in Chapter 6 to define the addition and multiplication of natural numbers, transfinite recursion may be used to give alternative definitions for the addition and multiplication of ordinal numbers.

10.2 Definition For any arbitrary $\alpha \in \mathrm{OR}$, we define a function $\sigma_\alpha : \mathrm{OR} \to \mathrm{OR}$ as follows:

a) $\sigma_\alpha(0) = \alpha$,

b) $\sigma_\alpha(\beta^+) = [\sigma_\alpha(\beta)]^+$,

c) $\sigma_\alpha(\beta) = \sup\{\sigma_\alpha(\gamma) \mid \gamma < \beta\}$ if β is a limit ordinal.

Theorem 10.1 guarantees the existence of a unique function σ_α satisfying (a), (b), (c).

10.3 Theorem For arbitrary ordinals α, β, $\sigma_\alpha(\beta) = \alpha + \beta$.

Proof. If $\beta = 0$, then $\sigma_\alpha(0) = \alpha = \alpha + 0$. By induction, let us suppose now that the theorem holds for all $\gamma < \beta$. If β is a limit ordinal, then

$$\begin{aligned} \sigma_\alpha(\beta) &= \sup\{\sigma_\alpha(\gamma) \mid \gamma < \beta\} && \text{by 10.2(c)} \\ &= \sup\{\alpha + \gamma \mid \gamma < \beta\} && \text{by the hypothesis of induction} \\ &= \alpha + \beta && \text{by Exercise 12(b), Exercise Set 9.3.} \end{aligned}$$

If β is a nonlimit ordinal, $\beta = \delta^+$, then

$$\begin{aligned} \sigma_\alpha(\beta) = \sigma_\alpha(\delta^+) &= [\sigma_\alpha(\delta)]^+ && \text{by 10.2(b)} \\ &= (\alpha + \delta)^+ && \text{by the hypothesis of induction} \\ &= \alpha + \delta^+ && \text{immediate consequence of 9.9(i)} \\ &= \alpha + \beta. \ ■ \end{aligned}$$

Theorem 10.3 tells us that if we define the addition of ordinal numbers by

10.4 $$\alpha + \beta = \sigma_\alpha(\beta), \qquad \sigma_\alpha \text{ given by 10.2,}$$

then 10.4 is equivalent to Definition 9.8.

10.4 may also be written in the following form:

10.5 a) $\alpha + 0 = \alpha$,

b) $\alpha + \beta^+ = (\alpha + \beta)^+$,

c) $\alpha + \beta = \sup\{\alpha + \gamma \mid \gamma < \beta\}$ if β is a limit ordinal.

Since 10.5 is equivalent to Definition 9.8, we will henceforth consider 10.5 to define the addition of ordinal numbers.

10.6 Definition For an arbitrary $\alpha \in \mathrm{OR}$, we define a function $\pi_\alpha\colon \mathrm{OR} \to \mathrm{OR}$ as follows:

a) $\pi_\alpha(0) = 0$,

b) $\pi_\alpha(\beta^+) = \pi_\alpha(\beta) + \alpha$,

c) $\pi_\alpha(\beta) = \sup\{\pi_\alpha(\gamma) \mid \gamma < \beta\}$ if β is a limit ordinal.

Theorem 10.1 guarantees the existence of a function π_α satisfying the conditions (a), (b), and (c).

10.7 Theorem For arbitrary ordinals α, β, $\pi_\alpha(\beta) = \alpha\beta$.

Proof. If $\beta = 0$, then $\pi_\alpha(0) = 0 = \alpha 0$. By induction, let us suppose that the theorem holds for every $\gamma < \beta$; that is, $\pi_\alpha(\gamma) = \alpha\gamma$, $\forall \gamma < \beta$. If β is a limit ordinal then

$$\begin{aligned} \pi_\alpha(\beta) &= \sup\{\pi_\alpha(\gamma) \mid \gamma < \beta\} && \text{by 10.6(c)} \\ &= \sup\{\alpha\gamma \mid \gamma < \beta\} && \text{by the hypothesis of induction} \\ &= \alpha\beta && \text{by Exercise 13, Exercise Set 9.3.} \end{aligned}$$

If β is a nonlimit ordinal, $\beta = \delta^+$, then

$$\begin{aligned} \pi_\alpha(\beta) = \pi_\alpha(\delta^+) &= \pi_\alpha(\delta) + \alpha && \text{by 10.6(b)} \\ &= \alpha\delta + \alpha && \text{by the hypothesis of induction} \\ &= \alpha(\delta + 1) && \text{by 9.9(iii)} \\ &= \alpha\beta. \ \blacksquare \end{aligned}$$

Theorem 10.7 tells us that if we define the multiplication of ordinal numbers by

10.8 $$\alpha\beta = \pi_\alpha(\beta), \quad \text{where } \pi_\alpha \text{ is given by 10.6,}$$

then 10.8 is equivalent to Definition 9.8.

10.8 may also be written in the following form:

10.9 a) $\alpha 0 = 0$,

b) $\alpha\beta^+ = \alpha\beta + \alpha$,

c) $\alpha\beta = \sup\{\alpha\gamma \mid \gamma < \beta\}$, if β is a limit ordinal.

Since 10.9 is equivalent to 9.8, we will henceforth consider 10.9 to be the definition of ordinal multiplication.

2 PROPERTIES OF ORDINAL EXPONENTIATION

10.5 and 10.9 provide us with an alternative way of defining the addition and multiplication of ordinal numbers, using transfinite recursion. We will use this new (and in many ways, more convenient) approach to define ordinal exponentiation.

10.10 Definition If $\alpha \neq 0$ is any ordinal, we define a function $\eta_\alpha : \mathrm{OR} \to \mathrm{OR}$ as follows:

a) $\eta_\alpha(0) = 1$,

b) $\eta_\alpha(\beta^+) = \eta_\alpha(\beta)\alpha$,

c) $\eta_\alpha(\beta) = \sup\{\eta_\alpha(\gamma) \mid \gamma < \beta\}$ if β is a limit ordinal.

Theorem 10.1 guarantees the existence of a function $\eta_\alpha : \mathrm{OR} \to \mathrm{OR}$ which satisfies conditions (a), (b), and (c) above.

10.11 Definition We define ordinal exponentiation as follows: If α and β are arbitrary ordinals, we let

$$\alpha^\beta = \eta_\alpha(\beta) \qquad \text{if} \quad \alpha \neq 0,$$

and $0^\beta = 0$.

In view of 10.10, 10.11 may therefore be written as follows:

10.12 a) $\alpha^0 = 1$,

b) $\alpha^{\beta+} = (\alpha^\beta)\alpha$,

c) $\alpha^\beta = \sup\{\alpha^\gamma \mid \gamma < \beta\}$, if β is a limit ordinal,

d) $0^\beta = 0$.

We will now develop the fundamental properties of ordinal exponentiation.

10.13 Lemma If γ is a limit ordinal and $\varepsilon < \alpha^\gamma$, then $\exists \pi < \gamma \ni \varepsilon < \alpha^\pi$.

Proof. Suppose, on the contrary, that $\forall \pi < \gamma$, $\alpha^\pi \leqslant \varepsilon$; this means that ε is an upper bound of the set $\{\alpha^\pi \mid \pi < \gamma\}$; but α^γ is the sup of this same set, so $\alpha^\gamma \leqslant \varepsilon$, which contradicts our assumption that $\varepsilon < \alpha^\gamma$. Thus for some $\pi < \gamma$, $\varepsilon < \alpha^\pi$. ■

10.14 Theorem For any ordinal numbers $\alpha > 1$, β, γ,

i) $\beta < \gamma \Rightarrow \alpha^\beta < \alpha^\gamma$,

ii) $\alpha^\gamma < \alpha^\beta \Rightarrow \gamma < \beta$.

Proof

i) The proof is by induction on γ. If $\gamma = 0$, the condition is satisfied vacuously. Now suppose that (i) holds $\forall\delta < \gamma$, that is,

$$\beta < \delta \Rightarrow \alpha^\beta < \alpha^\delta \qquad \text{for every} \quad \delta < \gamma.$$

Suppose first that γ is a limit ordinal. If $\beta < \gamma$, then $\beta + 1 < \gamma$, and by the hypothesis of induction we have

$$\beta < \beta + 1 \Rightarrow \alpha^\beta < \alpha^{\beta+1};$$

but $\alpha^\gamma = \sup\{\alpha^\beta \mid \beta < \gamma\}$, so $\alpha^{\beta+1} \leqslant \alpha^\gamma$; thus $\alpha^\beta < \alpha^\gamma$.

Now suppose that γ is a nonlimit ordinal, $\gamma = \delta + 1$. If $\beta < \gamma$, then $\beta = \delta$ or $\beta < \delta$. If $\beta = \delta$, then we have

$$\begin{aligned} \alpha^\gamma = \alpha^{\delta+1} &= \alpha^\delta\alpha && \text{by 10.12(b)} \\ &> \alpha^\delta && \text{by 9.16(v) (note that } \alpha > 1) \\ &= \alpha^\beta. \end{aligned}$$

If $\beta < \delta$, then by the hypothesis of induction, $\alpha^\beta < \alpha^\delta$, so we have

$$\begin{aligned} \alpha^\beta < \alpha^\delta &< \alpha^\delta\alpha && \text{by 9.16(v)} \\ &= \alpha^{\delta+1} = \alpha^\gamma && \text{by 10.12(b).} \end{aligned}$$

ii) $\alpha^\gamma \leqslant \alpha^\beta \Rightarrow \gamma \leqslant \beta$ is the contrapositive of (i). Now suppose $\alpha^\gamma < \alpha^\beta$; if $\gamma = \beta$, then $\alpha^\gamma = \alpha^\beta$, hence $\gamma < \beta$. ■

10.15 Theorem For any ordinal numbers α, β, and γ,

i) $\alpha \leqslant \beta \Rightarrow \alpha^\gamma \leqslant \beta^\gamma$,

ii) $\beta^\gamma < \alpha^\gamma \Rightarrow \beta < \alpha$.

Proof

i) The proof is by induction on γ. If $\gamma = 0$, the condition is satisfied trivially. Now suppose that (i) holds $\forall\delta < \gamma$, that is,

$$\alpha \leqslant \beta \Rightarrow \alpha^\delta \leqslant \beta^\delta \qquad \text{for every } \delta < \gamma.$$

Suppose first that γ is a nonlimit ordinal, $\gamma = \delta + 1$. We assume $\alpha \leqslant \beta$ and, by the hypothesis of induction, $\alpha^\delta \leqslant \beta^\delta$. Thus, by 10.12(b) and 9.16(v) and (vii), we have

$$\alpha^\gamma = \alpha^{\delta+1} = \alpha^\delta\alpha \leqslant \alpha^\delta\beta \leqslant \beta^\delta\beta = \beta^{\delta+1} = \beta^\gamma.$$

Next, suppose that γ is a limit ordinal; then $\alpha^\gamma = \sup\{\alpha^\delta \mid \delta < \gamma\}$. If $\delta < \gamma$, then by the hypothesis of induction $\alpha^\delta \leqslant \beta^\delta$; but $\beta^\delta \leqslant \beta^\gamma$ because $\beta^\gamma = \sup\{\beta^\delta \mid \delta < \gamma\}$, hence $\alpha^\delta \leqslant \beta^\gamma$ for every $\delta < \gamma$. It follows that β^γ is an upper bound of $\{\alpha^\delta \mid \delta < \gamma\}$, hence $\alpha^\gamma \leqslant \beta^\gamma$.

ii) This is simply the contrapositive of (i). ■

10.16 Theorem $\alpha^\beta\alpha^\gamma = \alpha^{\beta+\gamma}$ for any ordinal numbers α, β, and γ.

Proof. The proof is by induction on γ; the theorem holds trivially if $\gamma = 0$, hence we assume that $\alpha^\beta\alpha^\delta = \alpha^{\beta+\delta}$ for every ordinal $\delta < \gamma$.

i) Let us suppose first that γ is a nonlimit ordinal, $\gamma = \delta + 1$; then

$$\alpha^{\beta+\gamma} = \alpha^{\beta+\delta+1} = \alpha^{\beta+\delta}\alpha = \alpha^\beta\alpha^\delta\alpha = \alpha^\beta\alpha^{\delta+1} = \alpha^\beta\alpha^\gamma.$$

ii) Now let us suppose that γ is a limit ordinal; we shall prove the two inequalities (a) $\alpha^{\beta+\gamma} \leqslant \alpha^\beta\alpha^\gamma$ and (b) $\alpha^\beta\alpha^\gamma \leqslant \alpha^{\beta+\gamma}$.

a) If γ is a limit ordinal, then clearly $\beta + \gamma$ is a limit ordinal, hence by 10.12(c),

$$\alpha^{\beta+\gamma} = \sup\{\alpha^\delta \mid \delta < \beta + \gamma\}.$$

Now if $\delta < \beta + \gamma$, then either $\delta \leqslant \beta$, or if $\delta > \beta$, then by 9.15, $\delta = \beta + \rho$ for some $\rho < \gamma$ [$\rho < \gamma$ because $\delta = \beta + \rho < \beta + \gamma \Rightarrow \rho < \gamma$ by 9.16(ii)]. In the first case, namely $\delta \leqslant \beta$, it follows by 10.14(i) and 9.16(v) that $\alpha^\delta \leqslant \alpha^\beta \leqslant \alpha^\beta\alpha^\gamma$. (We assume that $\alpha \neq 0$, hence $1 \leqslant \alpha^\gamma$; if $\alpha = 0$, then 10.16 holds trivially.) In the second case, namely $\delta = \beta + \rho$ where $\rho < \gamma$, it follows by 10.14(i) that $\alpha^\rho < \alpha^\gamma$, hence by the hypothesis of induction and 9.16(v),

$$\alpha^\delta = \alpha^{\beta+\rho} = \alpha^\beta\alpha^\rho < \alpha^\beta\alpha^\gamma.$$

Thus, in either of the two cases, $\alpha^\delta \leqslant \alpha^\beta\alpha^\gamma$ for every $\delta < \beta + \gamma$, so $\alpha^\beta\alpha^\gamma$ is an upper bound of $\{\alpha^\delta \mid \delta < \beta + \gamma\}$, hence $\alpha^{\beta+\gamma} \leqslant \alpha^\beta\alpha^\gamma$.

b) We are assuming that γ is a limit ordinal; it follows very easily (see Exercise 1(a) at the end of this section) that α^γ is a limit ordinal, hence by 10.9(c),

$$\alpha^\beta\alpha^\gamma = \sup\{\alpha^\beta\varepsilon \mid \varepsilon < \alpha^\gamma\}.$$

Now if $\varepsilon < \alpha^\gamma$, then by 10.13, $\exists\pi < \gamma \ni \varepsilon < \alpha^\pi$; hence by 9.16(v), 10.14(i), and the hypothesis of induction,

$$\alpha^\beta\varepsilon < \alpha^\beta\alpha^\pi = \alpha^{\beta+\pi} < \alpha^{\beta+\gamma}.$$

Thus $\alpha^{\beta+\gamma}$ is an upper bound of $\{\beta^\beta\varepsilon \mid \varepsilon < \alpha^\gamma\}$, hence $\alpha^\beta\alpha^\gamma \leqslant \alpha^{\beta+\gamma}$. ■

10.17 Theorem $(\alpha^\beta)^\gamma = \alpha^{\beta\gamma}$ for any ordinals α, β, and γ.

Proof. The proof is by induction on γ. If $\gamma = 0$, the theorem follows trivially from 10.12(a). Let us assume, then, that $(\alpha^\beta)^\delta = \alpha^{\beta\delta}$ for every $\delta < \gamma$.

i) Suppose first that γ is a nonlimit ordinal, $\gamma = \delta + 1$; then by 10.12(b), the hypothesis of induction, and 10.16, we have

$$(\alpha^\beta)^\gamma = (\alpha^\beta)^{\delta+1} = (\alpha^\beta)^\delta\alpha^\beta = \alpha^{\beta\delta}\alpha^\beta = \alpha^{\beta\delta+\beta} = \alpha^{\beta(\delta+1)} = \alpha^{\beta\gamma}.$$

ii) Next, we shall suppose that γ is a limit ordinal and we will prove the two inequalities (a) $(\alpha^\beta)^\gamma \leqslant \alpha^{\beta\gamma}$ and (b) $\alpha^{\beta\gamma} \leqslant (\alpha^\beta)^\gamma$.

a)
$$\begin{aligned}(\alpha^\beta)^\gamma &= \sup\{(\alpha^\beta)^\delta \mid \delta < \gamma\} && \text{by 10.12(c)}\\ &= \sup\{\alpha^{\beta\delta} \mid \delta < \gamma\} && \text{by the hypothesis of induction.}\end{aligned}$$

But if $\delta < \gamma$, then $\beta\delta < \beta\gamma$, so by 10.14(i), $\alpha^{\beta\delta} < \alpha^{\beta\gamma}$; it follows that $\alpha^{\beta\gamma}$ is an upper bound of $\{\alpha^{\beta\delta} \mid \delta < \gamma\}$, so $(\alpha^\beta)^\gamma \leqslant \alpha^{\beta\gamma}$.

b) If γ is a limit ordinal, then (see Exercise 6, Exercise Set 9.3) $\beta\gamma$ is a limit ordinal; thus by 10.12(c),

$$\alpha^{\beta\gamma} = \sup\{\alpha^\delta \mid \delta < \beta\gamma\}.$$

Now if $\delta < \beta\gamma$, then by 9.18, $\delta = \beta\xi + \varepsilon$, where $\xi < \gamma$ and $\varepsilon < \beta$; thus

$$\delta = \beta\xi + \varepsilon < \beta\xi + \beta = \beta(\xi + 1).$$

Now $\xi < \gamma$ and γ is a limit ordinal, so $\xi + 1 < \gamma$; thus, by the hypothesis of induction and 10.14(i),

$$\alpha^{\beta(\xi+1)} = (\alpha^\beta)^{\xi+1} < (\alpha^\beta)^\gamma.$$

Thus, using 10.14(i) once again, $\alpha^\delta < \alpha^{\beta(\xi+1)} < (\alpha^\beta)^\gamma$. It follows that $(\alpha^\beta)^\gamma$ is an upper bound of $\{\alpha^\delta \mid \delta < \beta\gamma\}$, so $\alpha^{\beta\gamma} \leqslant (\alpha^\beta)^\gamma$. ■

Note that by 10.12(c), $2^\omega = \sup\{2^n \mid n < \omega\} = \omega$. We noted in the preceding chapter that the "usual" arithmetic laws do not *all* apply to transfinite ordinal numbers; in particular, the commutative law for multiplication does not hold, nor does the right distributive law. As we shall now see, the law $(\alpha\beta)^\gamma = \alpha^\gamma\beta^\gamma$ does not apply generally to ordinal numbers.

10.18 Example

i) $(2\cdot 2)^\omega = 4^\omega = \omega$ because $4^\omega = \sup\{4^n \mid n < \omega\} = \omega$.

ii) $2^\omega 2^\omega = \omega\omega = \omega^2$. Since $\omega > 1$, it follows by 9.16(v) that $\omega\omega > \omega 1 = \omega$; thus $(2\cdot 2)^\omega \neq 2^\omega 2^\omega$.

EXERCISES 10.2

1. Prove the following:

   a) If γ is a limit ordinal and $\alpha > 1$, then α^γ is a limit ordinal.

   b) If α is a limit ordinal and $\gamma \neq 0$, then α^γ is a limit ordinal.

2. Prove that for any ordinals $\alpha > 1$, β, and γ, $\alpha^\beta = \alpha^\gamma \Rightarrow \beta = \gamma$.

3. Use 10.5, 10.9, and 10.12 to prove that for any finite ordinal n,
 a) $n + \omega = \omega$, b) $n\omega = \omega$, c) $n^\omega = \omega$.
4. Use induction to prove that for every ordinal number β, $2^\beta \geqslant \beta$. Conclude that for every $\alpha > 0$ and β, $\alpha^\beta \geqslant \beta$.
5. Prove that if $\alpha > 1$ and $\beta \neq 0$, then $\alpha\beta \leqslant \alpha^\beta$.
6. Prove that if α is a limit ordinal and p and q are finite ordinals, then $(\alpha p)^q = \alpha^q p$. [*Hint:* Use Exercise 8, Exercise Set 9.3.]
7. Let a limit ordinal γ be called *simple* if it cannot be written $\gamma = \delta + \omega$ for any ordinal δ. Prove that γ is simple if and only if $\forall \varepsilon < \gamma$, there exists a limit ordinal $\lambda \ni \varepsilon < \lambda < \gamma$.
8. If α is a denumerable ordinal (that is, if $\#\alpha = \aleph_0$), use 9.8 to prove that $\alpha\omega$ is denumerable. Conclude that the ordinals $\omega, \omega^2, \ldots, \omega^n, \ldots$ (n finite) are all denumerable.
9. a) Let $\{\gamma_i \mid i \in I\}$ be a set of ordinals; prove that
$$\sup\{\gamma_i \mid i \in I\} = \bigcup_{i \in I} \gamma_i.$$
 [Use 9.25 and 9.26(ii).]
 b) Prove that if $\{\gamma_n \mid n \in \alpha\}$ is a set of denumerable ordinals, where α is denumerable, then $\sup\{\gamma_n \mid n \in \alpha\}$ is a denumerable ordinal. *Note:* It follows immediately from Exercises 2 and 9, Exercise Set 8.5 that
$$\#\bigcup_{i \in I} \gamma_i \leqslant \sum_{i \in I} (\#\gamma_i) \leqslant \aleph_0.$$
 c) Use the result of Exercise 8 above to prove that ω^ω is a denumerable ordinal. (Note that $\omega^\omega = \sup\{\omega^n \mid n \in \omega\}$.)
 d) Conclude similarly that ω^{ω^ω}, $\omega^{\omega^{\omega^\omega}}$, etc., are denumerable ordinals.
10. Prove that if α and β are denumerable ordinals, then α^β is a denumerable ordinal. [Use 9(ii) above.]

3 NORMAL FORM

It is a well-known fact of elementary number theory that every natural number n has a uniquely determined decimal representation. That is, given n, there exist unique natural numbers $k, m_0, m_1, \ldots, m_k$ (each $m_i < 10$) such that
$$n = m_k \cdot 10^k + \cdots + m_1 \cdot 10 + m_0.$$
The decimal representation of n is also called its representation with base 10; it can easily be shown that every natural number n also has a unique representation with base b, for any $b > 1$.

The idea of giving every number a representation with base b can easily be extended to ordinal numbers generally. In the sequel we will only consider base ω, but it should be clear to the reader that any other base will do.

10.19 Definition Let γ be an ordinal number; suppose there are nonzero natural numbers $a_1, \ldots, a_n$ and ordinals $\alpha_1 > \alpha_2 > \cdots > \alpha_n$ such that

10.20 $$\gamma = \omega^{\alpha_1}a_1 + \omega^{\alpha_2}a_2 + \cdots + \omega^{\alpha_n}a_n.$$

Then 10.20 is called a *normal form representation* of γ.

10.21 Theorem Every ordinal $\gamma \neq 0$ has a normal form representation.

Proof. The proof will be by induction on γ. If $\gamma = 1$, then $\gamma = \omega^0 1$ is a normal form representation of γ. Now assume the theorem is true $\forall \rho < \gamma$. Let $A = \{\mu \mid \omega^\mu > \gamma\}$; A is nonempty, as may easily be seen by using Exercise 4, Exercise Set 10.2. Thus A has a least element ν; $\omega^\nu > \gamma$. Suppose ν is a limit ordinal, $\omega^\nu = \sup\{\omega^\delta \mid \delta < \nu\}$. For each $\delta < \nu$, $\omega^\delta \leqslant \gamma$ because ν is the *least* element of A; thus γ is an upper bound of $\{\omega^\delta \mid \delta < \nu\}$, so $\omega^\nu \leqslant \gamma$. This is contrary to our choice of ν, hence $\nu = \alpha_1 + 1$ for some ordinal α_1.

By 9.19, $\gamma = \omega^{\alpha_1}\xi + \rho$, where $\rho < \omega^{\alpha_1}$; clearly $\xi < \omega$, for if $\xi \geqslant \omega$ then

$$\omega^{\alpha_1}\xi \geqslant \omega^{\alpha_1}\omega = \omega^\nu > \gamma,$$

which is impossible by 9.14(i). It follows that ξ is a natural number a_1, so $\gamma = \omega^{\alpha_1}a_1 + \rho$, where $\rho < \omega^{\alpha_1} \leqslant \gamma$. By the hypothesis of induction, there exist nonzero natural numbers $a_2, \ldots, a_n$ and ordinals $\alpha_2 > \cdots > \alpha_n$ such that

$$\rho = \omega^{\alpha_2}a_2 + \cdots + \omega^{\alpha_n}a_n;$$

clearly $\alpha_1 > \alpha_2$, for if $\alpha_1 \leqslant \alpha_2$, then

$$\omega^{\alpha_1} \leqslant \omega^{\alpha_2} \leqslant \omega^{\alpha_2}a_2 \leqslant \rho,$$

which is false because $\rho < \omega^{\alpha_1}$. Thus,

$$\gamma = \omega^{\alpha_1}a_1 + \omega^{\alpha_2}a_2 + \cdots + \omega^{\alpha_n}a_n$$

where $a_1, \ldots, a_n$ are nonzero natural numbers and $\alpha_1 > \cdots > \alpha_n$. ■

We will prove next that the normal form representation of γ is unique.

10.22 Lemma If $\gamma = \omega^{\alpha_1}a_1 + \cdots + \omega^{\alpha_n}a_n$ is a normal form representation of γ, then $\gamma < \omega^{\alpha_1+1}$.

Proof. By 10.19, $\alpha_i < \alpha_1$ for $i = 2, \ldots, n$; thus $\omega^{\alpha_i} < \omega^{\alpha_1}$ for $i = 2, \ldots, n$. Thus,

$$\begin{aligned}\gamma = \omega^{\alpha_1}a_1 + \omega^{\alpha_2}a_2 + \cdots + \omega^{\alpha_n}a_n &\leqslant \omega^{\alpha_1}a_1 + \omega^{\alpha_1}a_2 + \cdots + \omega^{\alpha_1}a_n \\ &= \omega^{\alpha_1}(a_1 + \cdots + a_n) < \omega^{\alpha_1}\omega = \omega^{\alpha_1+1}.\end{aligned}$$ ■

10.23 Theorem The normal form representation of any ordinal γ is unique.

Proof. Suppose $\gamma = \omega^{\alpha_1}a_1 + \cdots + \omega^{\alpha_n}a_n = \omega^{\beta_1}b_1 + \cdots + \omega^{\beta_m}b_m$, where $a_1, \ldots, a_n$, $b_1, \ldots, b_m$ are nonzero natural numbers, $\alpha_1 > \alpha_2 > \cdots > \alpha_n$ and $\beta_1 > \cdots > \beta_m$. Let us write

$$\rho_a = \omega^{\alpha_2}a_2 + \cdots + \omega^{\alpha_n}a_n \quad \text{and} \quad \rho_b = \omega^{\beta_2}b_2 + \cdots + \omega^{\beta_m}b_m;$$

thus,

$$\gamma = \omega^{\alpha_1}a_1 + \rho_a = \omega^{\beta_1}b_1 + \rho_b.$$

Suppose $\alpha_1 < \beta_1$; then $\alpha_1 + 1 \leqslant \beta_1$, so

$$\omega^{\alpha_1+1} \leqslant \omega^{\beta_1} \leqslant \omega^{\beta_1}b_1 \leqslant \gamma.$$

But by 10.22, $\gamma < \omega^{\alpha_1+1}$, so we have a contradiction. Analogously, we cannot have $\beta_1 < \alpha_1$, hence $\alpha_1 = \beta_1$. Thus,

$$\gamma = \omega^{\alpha_1}a_1 + \rho_a = \omega^{\alpha_1}b_1 + \rho_b.$$

Now suppose $a_1 < b_1$, hence $a_1 + 1 \leqslant b_1$; then

$$\omega^{\alpha_1}(a_1 + 1) \leqslant \omega^{\alpha_1}b_1,$$

that is,

$$\omega^{\alpha_1}a_1 + \omega^{\alpha_1} \leqslant \omega^{\alpha_1}b_1.$$

Thus

$$\omega^{\alpha_1}a_1 + \omega^{\alpha_1} + \rho_b \leqslant \omega^{\alpha_1}b_1 + \rho_b = \gamma.$$

This gives us $\omega^{\alpha_1}a_1 + \omega^{\alpha_1} + \rho_b \leqslant \omega^{\alpha_1}a_1 + \rho_a$, hence $\omega^{\alpha_1} + \rho_b \leqslant \rho_a$, so $\omega^{\alpha_1} \leqslant \rho_a$. But this is impossible, for $\alpha_2 < \alpha_1$, hence $\omega^{\alpha_2+1} \leqslant \omega^{\alpha_1}$, so by 10.22, $\rho_a < \omega^{\alpha_2+1} \leqslant \omega^{\alpha_1}$. Consequently, we cannot have $a_1 < b_1$; analogously, we cannot have $b_1 < a_1$, so $a_1 = b_1$.

Thus, $\gamma = \omega^{\alpha_1}a_1 + \rho_a = \omega^{\alpha_1}a_1 + \rho_b$, so $\rho_a = \rho_b$. By induction, we may now assume that the normal form representation of $\rho_a = \rho_b$ is unique, hence $\alpha_2 = \beta_2, \ldots, \alpha_n = \beta_n, a_2 = b_2, \ldots, a_n = b_n$. ■

The theorem which follows makes it easy to add and multiply ordinal numbers when they are written in normal form.

10.24 Theorem

i) If $\alpha < \beta$, then $\omega^\alpha a + \omega^\beta b = \omega^\beta b$.

ii) If $\gamma = \omega^{\alpha_1}a_1 + \cdots + \omega^{\alpha_n}a_n$ is the normal form representation of γ and if $\beta \neq 0$, then $\gamma\omega^\beta = \omega^{\alpha_1+\beta}$.

iii) If $\gamma = \omega^{\alpha_1}a_1 + \cdots + \omega^{\alpha_n}a_n$ is the normal form of γ and if b is finite, then $\gamma b = \omega^{\alpha_1}a_1 b + \omega^{\alpha_2}a_2 + \cdots + \omega^{\alpha_n}a_n$.

Proof

i) If $\alpha < \beta$, then by 9.15, $\beta = \alpha + \delta$ for some $\delta > 0$. Thus,

$$\omega^\alpha a + \omega^\beta b = \omega^\alpha a + \omega^{\alpha+\delta}b = \omega^\alpha a + \omega^\alpha\omega^\delta b = \omega^\alpha(a + \omega^\delta b).$$

Since a is finite, it is clear that $a + \omega^\delta b = \omega^\delta b$ (see, for example, Exercise 3, Exercise Set 10.2). Thus

$$\omega^\alpha a + \omega^\beta b = \omega^\alpha(a + \omega^\delta b) = \omega^\alpha(\omega^\delta b) = \omega^{\alpha+\delta} b = \omega^\beta b.$$

ii) It can be proven very easily that $n\omega = \omega$ for every $n \in \omega$ (see, for example, Exercise 3, Exercise Set 10.2). It follows that $n\omega^\beta = \omega^\beta$ for every $\beta > 0$ and $n \in \omega$. Indeed, for $\beta = 1$ this has just been given: if $\beta > 1$, then by 9.14(i), $\beta = 1 + \delta$ for some $\delta > 0$, so we have

$$n\omega^\beta = n\omega^{1+\delta} = n(\omega\omega^\delta) = (n\omega)\omega^\delta = \omega\omega^\delta = \omega^{1+\delta} = \omega^\beta.$$

Thus

$$\begin{aligned}
\gamma\omega^\beta &= (\omega^{\alpha_1}a_1 + \cdots + \omega^{\alpha_n}a_n)\omega^\beta \\
&\leqslant (\omega^{\alpha_1}a_1 + \cdots + \omega^{\alpha_1}a_n)\omega^\beta \\
&= \omega^{\alpha_1}(a_1 + \cdots + a_n)\omega^\beta \\
&= \omega^{\alpha_1}m\omega^\beta && \text{where } m = a_1 + \cdots + a_n \text{ is finite} \\
&= \omega^{\alpha_1}\omega^\beta && \text{because } m\omega^\beta = \omega^\beta, \text{ as above} \\
&= \omega^{\alpha_1+\beta}.
\end{aligned}$$

On the other hand, $\omega^{\alpha_1} \leqslant \omega^{\alpha_1}a_1 \leqslant \gamma$, hence $\omega^{\alpha_1+\beta} = \omega^{\alpha_1}\omega^\beta \leqslant \gamma\omega^\beta$.

iii) The proof is by finite induction on b. If $b = 1$, there is nothing to prove. Now suppose (iii) holds for b, and let us prove it for $b + 1$. We have

$$\begin{aligned}
(\omega^{\alpha_1}a_1 + \cdots + \omega^{\alpha_n}a_n)(b + 1) &= (\omega^{\alpha_1}a_1 + \cdots + \omega^{\alpha_n}a_n)b \\
&\quad + (\omega^{\alpha_1}a_1 + \cdots + \omega^{\alpha_n}a_n) \\
&= (\omega^{\alpha_1}a_1 b + \omega^{\alpha_2}a_2 + \cdots + \omega^{\alpha_n}a_n) \\
&\quad + (\omega^{\alpha_1}a_1 + \cdots + \omega^{\alpha_n}a_n)
\end{aligned}$$

by the hypothesis of induction. The reader should note that in the above sum, the terms $\omega^{\alpha_2}a_2, \ldots, \omega^{\alpha_n}a_n$ all precede the term $\omega^{\alpha_1}a_1$, and that $\alpha_1 > \alpha_2, \ldots, \alpha_1 > \alpha_n$; thus, by 10.24(i), they disappear from the sum. Thus,

$$\begin{aligned}
(\omega^{\alpha_1}a_1 + \cdots + \omega^{\alpha_n}a_n)(b + 1) &= \omega^{\alpha_1}a_1 b + \omega^{\alpha_1}a_1 + \cdots + \omega^{\alpha_n}a_n \\
&= \omega^{\alpha_1}a_1(b + 1) + \omega^{\alpha_2}a_2 + \cdots + \omega^{\alpha_n}a_n. \blacksquare
\end{aligned}$$

When adding or multiplying ordinal numbers in normal form, it is also useful to remember that if β is an infinite ordinal and n is finite, then $n + \beta = \beta$ (see, for example, Exercise 3, Exercise Set 10.2).

10.25 Example Let

$$\alpha = \omega^{\omega 2}8 + \omega^{\omega+1}2 + \omega^3 4 \quad \text{and} \quad \beta = \omega^{\omega+3}7 + \omega^{10}5 + 2.$$

We shall form the sums $\alpha + \beta$ and $\beta + \alpha$ and the products $\alpha\beta$ and $\beta\alpha$.

$$\begin{aligned}\alpha + \beta &= \omega^{\omega 2}8 + \omega^{\omega+1}2 + \omega^3 4 + \omega^{\omega+3}7 + \omega^{10}5 + 2\\ &= \omega^{\omega 2}8 + \omega^{\omega+3}7 + \omega^{10}5 + 2.\end{aligned}$$

Note that the terms $\omega^{\omega+1}2$ and $\omega^3 4$ precede $\omega^{\omega+3}7$, so by 10.24(i) they disappear from the sum.

$$\begin{aligned}\beta + \alpha &= \omega^{\omega+3}7 + \omega^{10}5 + 2 + \omega^{\omega 2}8 + \omega^{\omega+1}2 + \omega^3 4\\ &= \omega^{\omega 2}8 + \omega^{\omega+1}2 + \omega^3 4 = \alpha.\end{aligned}$$

$$\begin{aligned}\alpha\beta &= \alpha\omega^{\omega+3}7 + \alpha\omega^{10}5 + \alpha 2 && \text{by 9.9(iii)}\\ &= \omega^{\omega 2+\omega+3}7 + \omega^{\omega 2+10}5 + (\omega^{\omega 2}16 + \omega^{\omega+1}2 + \omega^3 4) && \text{by 10.24(ii)–(iii)}\\ &= \omega^{\omega 3+3}7 + \omega^{\omega 2+10}5 + \omega^{\omega 2}16 + \omega^{\omega+1}2 + \omega^3 4\end{aligned}$$

$$\begin{aligned}\beta\alpha &= \beta\omega^{\omega 2}8 + \beta\omega^{\omega+1}2 + \beta\omega^3 4 && \text{by 9.9(iii)}\\ &= \omega^{\omega+3+\omega 2}8 + \omega^{\omega+3+\omega+1}2 + \omega^{\omega+3+3}4 && \text{by 10.24(ii)}\\ &= \omega^{\omega 3}8 + \omega^{\omega 2+1}2 + \omega^{\omega+6}4. \text{ (Note that } 3 + \omega 2 = \omega 2 \text{ and } 3 + \omega = \omega.)\end{aligned}$$

EXERCISES 10.3

1. In each of the following, compute $\alpha + \beta$, $\beta + \alpha$, $\alpha\beta$, and $\beta\alpha$.

   a) $\alpha = \omega^{\omega 3}2 + \omega^{\omega}4 + \omega^{10}5$; $\beta = \omega^{\omega+1}7 + \omega^2 9 + 14$.
   b) $\alpha = \omega^{\omega^{\omega}}9 + \omega^{\omega}7 + \omega 2$; $\beta = \omega^{\omega 5}8 + \omega^7 2 + 1$.
   c) $\alpha = \omega^{\omega^{\omega}}22 + \omega^{\omega\,18}4 + 71$; $\beta = \omega^{\omega 2+3}12 + 100$.

2. Let γ be an ordinal number; prove that γ is irreducible (See Exercise 2, Exercise Set 9.3) if and only if $\gamma = \omega^{\beta}$ for some ordinal β.

In each of the following exercises, we will assume that

$$\gamma = \omega^{\alpha_1}a_1 + \cdots + \omega^{\alpha_n}a_n$$

is the normal form representation of γ.

3. Prove that γ is a limit ordinal if and only if $\alpha_n \neq 0$.

4. Let us define the *magnitude* of γ to be the ordinal α_1. Prove that $\alpha + \beta = \beta$ if and only if magnitude $\alpha <$ magnitude β.

5. Prove that $\omega\gamma = \gamma$ if and only if $\alpha_1, \ldots, \alpha_n$ are all infinite ordinals. Conclude that $\omega\gamma = \gamma$ if and only if $\gamma = \omega^{\omega}\beta$ for some ordinal β.

6. Let γ be a limit ordinal and let b be a finite ordinal. Use finite induction on b to prove that
$$\gamma^b = \omega^{\alpha_1 b}a_1 + \omega^{\alpha_1(b-1)}[\omega^{\alpha_2}a_2 + \cdots + \omega^{\alpha_n}a_n].$$

7. Use the result of Exercise 6 above to compute α^6 and β^{15}, where α and β are given in Exercise 1(a).

4 EPSILON NUMBERS

Cantor investigated the properties of an interesting class of limit ordinals which he called *epsilon numbers.* These numbers shed some light on the structure of the well-ordered class OR and have useful applications in analysis and elsewhere. We shall give a brief review of their properties in this section.

10.26 Definition Let α be an ordinal number; α is called an *epsilon number* if $\alpha = \omega^\alpha$.

It is immediate that every epsilon number is necessarily a limit ordinal. Now, the first question we are led to ask is: Are there any epsilon numbers? To answer this question, we first need a lemma.

10.27 Lemma Let $\{\beta_i \mid i \in I\}$ be a set of ordinals, and let $\beta = \sup\{\beta_i \mid i \in I\}$; then $\alpha^\beta = \sup\{\alpha^{\beta_i} \mid i \in I\}$.

Proof

i) Suppose that $\beta = \beta_i$ for some $i \in I$. Thus, $\forall j \in I, \beta_j \leqslant \beta_i$, hence $\alpha^{\beta_j} \leqslant \alpha^{\beta_i}$. It follows that

$$\alpha^\beta = \alpha^{\beta_i} = \sup\{\alpha^{\beta_j} \mid j \in I\}.$$

ii) Now suppose that $\forall i \in I, \beta \neq \beta_i$; β must be a limit ordinal, for if $\beta = \delta + 1$, then $\exists i \in I\ \beta_i > \delta$, so $\beta_i = \delta + 1 = \beta$, which is contrary to our assumption. Now $\alpha^\beta = \sup\{\alpha^\gamma \mid \gamma < \beta\}$; for each $i \in I$, $\beta_i < \beta$, hence $\alpha^{\beta_i} < \alpha^\beta$; thus, $\sup\{\alpha^{\beta_i} \mid i \in I\} \leqslant \alpha^\beta$. On the other hand, if $\gamma < \beta$, then $\beta_i > \gamma$ for some $i \in I$, hence $\alpha^\gamma < \alpha^{\beta_i} \leqslant \sup\{\alpha^{\beta_i} \mid i \in I\}$. Thus

$$\alpha^\beta = \sup\{\alpha^\gamma \mid \gamma < \beta\} \leqslant \sup\{\alpha^{\beta_i} \mid i \in I\}.$$

Consequently,

$$\alpha^\beta = \sup\{\alpha^{\beta_i} \mid i \in I\}. \blacksquare$$

We now return to the question: Are there any epsilon numbers? The answer is "yes," and this can easily be shown as follows:

We define a function $f_0 : \omega \to \text{OR}$ as follows:

$$f_0(0) = 1, \qquad f_0(n + 1) = \omega^{f_0(n)}.$$

The existence of f_0 is guaranteed by the finite recursion theorem, 6.8. Clearly, we have

$$f_0(0) = 1, f_0(1) = \omega, f_0(2) = \omega^\omega, f_0(3) = \omega^{\omega^\omega}, \quad \text{and so on.}$$

Now, let $\varepsilon_0 = \sup\{f(n) \mid n \in \omega\}$; we claim that ε_0 is an epsilon number. Indeed, by 10.27,

$$\omega^{\varepsilon_0} = \sup\{\omega^{f(n)} \mid n \in \omega\} = \sup\{f(n+1) \mid n \in \omega\} = \varepsilon_0.$$

Thus there is at least one epsilon number, namely ε_0; we can easily show, in fact, that ε_0 is the least epsilon number.

10.29 Theorem ε_0 is the least epsilon number.

Proof. If α is an epsilon number, then $f_0(0) = 1 \leqslant \alpha$ (for clearly 0 is not an epsilon number). Now suppose that $f(n) \leqslant \alpha$; then $f(n+1) = \omega^{f(n)} \leqslant \omega^\alpha = \alpha$. It follows by finite induction that $f(n) \leqslant \alpha$ for every $n \in \omega$; thus, $\varepsilon_0 \leqslant \alpha$. ■

Is ε_0 the only epsilon number or are there others? It is easy to answer this question, for the method we used to construct ε_0 may be used to construct infinitely many other epsilon numbers. Indeed, we have the following:

10.30 Theorem If α is any ordinal number, let f_α be the function defined inductively by the pair of conditions

$$f_\alpha(0) = \alpha + 1,$$
$$f_\alpha(n+1) = \omega^{f_\alpha(n)}$$

and let $S(\alpha) = \sup\{f_\alpha(n) \mid n \in \omega\}$. Then $S(\alpha)$ is an epsilon number; furthermore, $S(\alpha)$ is the least epsilon number greater than α.

Proof. By 10.27,

$$\omega^{S(\alpha)} = \sup\{\omega^{f_\alpha(n)} \mid n \in \omega\} = \sup\{f_\alpha(n+1) \mid n \in \omega\} = S(\alpha).$$

Clearly $\alpha < S(\alpha)$. Now let γ be an epsilon number such that $\alpha < \gamma$; then $f_\alpha(0) = \alpha + 1 \leqslant \gamma$. Furthermore, assuming that $f_\alpha(n) \leqslant \gamma$, we have

$$f_\alpha(n+1) = \omega^{f_\alpha(n)} \leqslant \omega^\gamma = \gamma.$$

It follows, by finite induction, that $f_\alpha(n) \leqslant \gamma$ for every $n \in \omega$, hence $S(\alpha) \leqslant \gamma$. Thus $S(\alpha)$ is the least epsilon number greater than α. ■

It is easy to see that ε_0 is $S(0)$; thus, the first few epsilon numbers are $S(0)$, $S(1)$, $S(2)$, $S(3)$, etc. The next question we are led to ask is: Are there any epsilon numbers greater than all the $S(n)$, $(n \in \omega)$? The question is easily answered in the following theorem.

10.31 Theorem Let $\{\alpha_i \mid i \in I\}$ be a set of epsilon numbers; if $\beta = \sup\{\alpha_i \mid i \in I\}$, then β is an epsilon number.

Proof. By 10.27,

$$\omega^{\beta} = \sup\{\omega^{\alpha_i} \mid i \in I\} = \sup\{\alpha_i \mid i \in I\} = \beta. \blacksquare$$

Let ε: OR → OR be the function defined recursively as follows [we agree to write ε_i for $\varepsilon(i)$]:

10.32

$$\begin{aligned} \varepsilon_0 &= S(0), \\ \varepsilon_{\alpha+1} &= S(\varepsilon_\alpha), \quad \forall \alpha \in \mathrm{OR} \\ \varepsilon_\beta &= \sup\{\varepsilon_\gamma \mid \gamma < \beta\} \quad \text{if } \beta \text{ is a limit ordinal.} \end{aligned}$$

10.33 Theorem ε is an isomorphism from OR to the class of all the epsilon numbers.

Proof. To prove that ε_β is an epsilon number for every $\beta \in$ OR, we argue by induction:

i) We have already seen that ε_0 is an epsilon number.
ii) Now assume that ε_α is an epsilon number for every $\alpha < \beta$. If β is a nonlimit ordinal, $\beta = \delta + 1$, then

$$\varepsilon_\beta = \varepsilon_{\delta+1} = S(\varepsilon_\delta)$$

is an epsilon number by 10.30. If β is a limit ordinal, then

$$\varepsilon_\beta = \sup\{\varepsilon_\gamma \mid \gamma < \beta\}$$

is an epsilon number by 10.31. Thus, $\forall \beta \in$ OR, ε_β is an epsilon number. It is immediate that ε is a strictly increasing function, hence it is injective. To show that the range of ε is the class E of the epsilon numbers, let $\delta \in E$ and let ε_γ be the least ε_ζ such that $\delta < \varepsilon_\zeta$. Now γ cannot be a limit ordinal, for if it is, then by 10.32,

$$\varepsilon_\gamma = \sup\{\varepsilon_\pi \mid \pi < \gamma\}$$

hence (because $\delta < \varepsilon_\gamma$) $\exists \pi < \gamma \ni \delta < \varepsilon_\pi$, which contradicts our choice of γ. Thus, γ is a nonlimit ordinal, $\gamma = \pi + 1$; by the choice of γ, $\varepsilon_\pi \leqslant \delta < \varepsilon_{\pi+1}$, so by 10.30, $\delta = \varepsilon_\pi$. The fact that ε is an isomorphism follows now by 4.48. $\blacksquare$

It turns out, then, that there are "as many" epsilon numbers as there are ordinals. The class of the epsilon numbers is

$$\varepsilon_0, \varepsilon_1, \varepsilon_2, \ldots, \varepsilon_\omega, \varepsilon_{\omega+1}, \ldots, \varepsilon_{\omega 2}, \ldots, \varepsilon_{\omega 3}, \ldots, \varepsilon_{\omega\omega}, \ldots, \varepsilon_\alpha, \ldots,$$

as α ranges over all the ordinals.

A few easy arithmetic rules simplify computations which involve epsilon numbers. They are given in the next theorem.

10.34 Theorem Let ε designate an arbitrary epsilon number. Then
i) If $\alpha < \varepsilon$, then $\alpha + \varepsilon = \varepsilon$.
ii) If $\alpha < \varepsilon$, then $\alpha\,\varepsilon = \varepsilon$.
iii) If $\alpha < \varepsilon$, then $\alpha^\varepsilon = \varepsilon$.

Proof. Let $\alpha < \varepsilon$, and write α in normal form:

$$\alpha = \omega^{\alpha_1}a_1 + \cdots + \omega^{\alpha_n}a_n.$$

i) $\alpha + \varepsilon = \alpha + \omega^\varepsilon = \omega^{\alpha_1}a_1 + \cdots + \omega^{\alpha_n}a_n + \omega^\varepsilon$. Now $\alpha < \varepsilon$, hence for $i = 1, \ldots, n$, $\omega^{\alpha_i} < \varepsilon = \omega^\varepsilon$, so $\alpha_i < \varepsilon$. It follows by 10.24(i) that

$$\omega^{\alpha_1}a_1 + \cdots + \omega^{\alpha_n}a_n + \omega^\varepsilon = \omega^\varepsilon = \varepsilon, \qquad \text{so } \alpha + \varepsilon = \varepsilon.$$

ii)
$$\begin{aligned} \alpha\varepsilon = \alpha\omega^\varepsilon &= (\omega^{\alpha_1}a_1 + \cdots + \omega^{\alpha_n}a_n)\omega^\varepsilon \\ &= \omega^{\alpha_1+\varepsilon} && \text{by 10.24(ii)} \\ &= \omega^\varepsilon && \text{by 10.34(i)} \\ &= \varepsilon. \end{aligned}$$

iii)
$$\begin{aligned} \alpha = \omega^{\alpha_1}a_1 + \cdots + \omega^{\alpha_n}a_n &\leqslant \omega^{\alpha_1}a_1 + \cdots + \omega^{\alpha_1}a_n \\ &= \omega^{\alpha_1}(a_1 + \cdots + a_n) < \omega^{\alpha_1}\omega = \omega^{\alpha_1+1}. \end{aligned}$$

Thus $\alpha^\varepsilon \leqslant (\omega^{\alpha_1+1})^\varepsilon = \omega^{(\alpha_1+1)\varepsilon}$. Since every epsilon number is a limit ordinal, $\alpha_1 < \varepsilon \Rightarrow \alpha_1 + 1 < \varepsilon$, so by 10.34(ii), $(\alpha_1 + 1)\varepsilon = \varepsilon$. Thus, finally,

$$\omega^{(\alpha_1+1)\varepsilon} = \omega^\varepsilon = \varepsilon, \quad \text{so } \alpha^\varepsilon \leqslant \varepsilon.$$

But we always have $\alpha^\varepsilon \geqslant \varepsilon$ (see Exercise 4, Exercise Set 10.2), so $\alpha^\varepsilon = \varepsilon$. ■

It is worth noting that α is an epsilon number if and only if α satisfies the following condition:

10.35 If $\beta < \alpha$ and $\gamma < \alpha$, then $\beta^\gamma < \alpha$.

The proof of this statement is left as an exercise for the reader (Exercise 4 below).

EXERCISES 10.4

1. Prove that if α is an epsilon number, then α is a limit ordinal.
2. If α is an epsilon number, prove that $\beta^{\omega^\alpha} = \beta^{\omega\alpha}$ for every ordinal number β.
3. Prove that if α is an infinite ordinal and $\alpha^\beta = \beta$, then β is an epsilon number.
4. a) Prove that if α $(\alpha > \omega)$ satisfies 10.35, then α is a limit ordinal.

b) Prove that α ($\alpha > \omega$) is an epsilon number if and only if α satisfies 10.35.

5. Prove that α ($\alpha > \omega$) is an epsilon number if and only if $2^\alpha = \alpha$.

5 INACCESSIBLE ORDINALS AND CARDINALS

Inaccessible ordinals and cardinals play an important role in current investigations on the axiomatic foundations of set theory. They also have applications in functional analysis, topology, algebra, mathematical logic and other areas of advanced mathematics. In this section we will introduce them and give a few of their basic properties.

If α is any ordinal number, let $Z(\alpha)$ designate the class of all the ordinals which are equipotent with α; the least element of $Z(\alpha)$ is called the *initial ordinal belonging to* α, and is denoted by $\text{Io}(\alpha)$. It is immediate that

10.36 $\forall \alpha \in \text{OR}, \text{Io}(\alpha) \leqslant \alpha,$

and

10.37 $\text{Io}(\alpha)$ is the largest initial ordinal less than α.

(To prove 10.37, note that if $\text{Io}(\alpha) < \gamma \leqslant \alpha$, then $\text{Io}(\alpha) \subseteq \gamma \subseteq \alpha$, hence $\text{Io}(\alpha) \approx \alpha \approx \gamma$.) We have seen that the class of all the initial ordinals satisfies conditions K1 and K2 of the axiom of cardinality, hence we are justified in making the following definition (see 9.33):

By a cardinal number we mean an initial ordinal; thus, the class CD of all the cardinal numbers is the class of all the initial ordinals.

We have noted that if α is any ordinal number, it is common practice to write $\aleph_\alpha = \omega_\alpha$, treating $\aleph_\alpha$ as a cardinal and ω_α as an ordinal. If α is any ordinal number, it is easy to see that

10.38 $$\text{Io}(\alpha) = \#\alpha.$$

It follows easily from 10.36, 10.37, and 10.38 that if A is any well-ordered set,

10.39 $$\circledcirc A \geqslant \omega_\alpha \quad \text{iff} \quad \#A \geqslant \aleph_\alpha.$$

(To prove 10.39, set $\gamma = \circledcirc A$; hence $\text{Io}(\gamma) = \#\gamma = \#A$; the simple details are left as an exercise for the reader.)

We will now begin the process of defining inaccessible ordinals and cardinals.

10.40 Definition Let A be a well-ordered class and let $B \subseteq A$; we say that B is a *cofinal subclass* of A if

$$\forall x \in A, \quad \exists y \in B \ni y > x.$$

10.41 Lemma If C is a cofinal subclass of B and B is a cofinal subclass of A, then C is a cofinal subclass of A.

Proof. Let $x \in A$; then $\exists y \in B \ni y > x$; hence $\exists z \in C \ni z > y$, so $z > x$. ■

10.42 Lemma Let α be an ordinal and let $B \subseteq \alpha$; B is a cofinal subset of α if and only if $\sup B = \alpha$.

Proof. The reader should note that by 9.26(ii) and 9.24, every element of an ordinal number is an ordinal number, and $\gamma < \alpha$ iff $\gamma \in \alpha$.

i) Let B be a cofinal subset of α; $\forall x \in B$, $x \in \alpha$, that is, $x < \alpha$, so α is an upper bound of B. Now if $\gamma < \alpha$, then $\gamma \in \alpha$, so by 10.40, $\exists \beta \in B \ni \beta > \gamma$, so γ is not an upper bound of B; this proves that α is the *least* upper bound of B.

ii) Suppose $\sup B = \alpha$; if $\gamma \in \alpha$, that is, $\gamma < \alpha$, then (because γ is *not* an upper bound of B) $\exists \beta \in B \ni \beta > \gamma$. Thus B is a cofinal subset of α. ■

10.43 Definition Let α be a limit ordinal; by the *cofinality* of α we mean the least ordinal β such that α has a cofinal subset similar to β. If β is the cofinality of α, we write $\beta = cf(\alpha)$.

10.44 Lemma If $\beta = cf(\alpha)$ for some $\alpha \in \text{OR}$, then $\beta = cf(\beta)$.

Proof. Let $\gamma = cf(\beta)$; now α has a cofinal subset similar to β and β has a cofinal subset similar to γ, hence by 10.41, α has a cofinal subset similar to γ. But β is the least ordinal γ such that α has a cofinal subset similar to γ, so $\beta \leqslant \gamma$; now $\gamma \leqslant \beta$ because γ is similar to a subset of β, so $\beta = \gamma$. ■

10.45 Theorem Let β be a limit ordinal; if $\beta = cf(\beta)$, then β is an initial ordinal.

Proof. Let $\omega_\alpha = \text{Io}(\beta)$, and let f be a bijective function $f: \omega_\alpha \to \beta$; let $A = \{\gamma \in \omega_\alpha \mid \forall \delta < \gamma, f(\delta) < f(\gamma)\}$. We will show first that $\bar{f}(A)$ is a cofinal subset of β. Indeed, if $v \in \beta$, then (because β is a limit ordinal and a limit ordinal has no greatest element), there are elements $\xi \in \omega_\alpha \ni f(\xi) > v$. (Remember that f is bijective!) Let π be the least such element; then $\forall \xi < \pi, f(\xi) \leqslant v < f(\pi)$, hence $\pi \in A$; this proves that $\forall v \in \beta$, $\exists \pi \in A \ni f(\pi) > v$; thus $\bar{f}(A)$ is a cofinal subset of β.

Because of the way we have defined A, it is easy to see that $f_{[A]}: A \to \bar{f}(A)$ is an isomorphism; thus, if $\delta = \circledcirc A$, then δ is similar to a cofinal subset of β, so $cf(\beta) \leqslant \delta$. But A is a subset of ω_α, so $\delta \leqslant \omega_\alpha$; thus, $\beta = cf(\beta) \leqslant \omega_\alpha$. But ω_α is the *least* ordinal equipotent with β, so $\omega_\alpha \leqslant \beta$. Thus $\omega_\alpha = \beta$, so β is an initial ordinal. ■

10.46 Definition A limit ordinal β is called *regular* if $cf(\beta) = \beta$.

It follows, by Lemma 10.44, that the class of the regular ordinals is the range of the function $cf\colon \mathrm{OR} \to \mathrm{OR}$. By Theorem 10.45, *every regular ordinal is an initial ordinal* ω_α. Thus, in particular, every regular ordinal is a cardinal.

10.47 Definition If ω_α is a regular ordinal, then $\aleph_\alpha$ is called a *regular cardinal.*

The significance of regular cardinals in set theory will become apparent once we have given an alternative definition for them. We begin with the following two lemmas:

10.48 Lemma Let $\{\gamma_i \mid i \in I\}$ be a set of ordinals, and let $\#\gamma_i = \aleph_{\delta_i}$ for each $i \in I$. If $\omega_\alpha = \sup\{\gamma_i \mid i \in I\}$, then $\alpha = \sup\{\delta_i \mid i \in I\}$.

Proof. If $\omega_\alpha = \sup\{\gamma_i \mid i \in I\}$, then, for each $i \in I$, $\gamma_i \leqslant \omega_\alpha$, hence by 10.39, $\aleph_{\delta_i} = \#\gamma_i \leqslant \aleph_\alpha$, so by 9.22, $\delta_i \leqslant \alpha$. Now suppose that for every $i \in I$, $\delta_i < \pi$; then $\forall i \in I$, $\#\gamma_i = \aleph_{\delta_i} < \aleph_\pi$, hence by 10.39 $\gamma_i < \omega_\pi$. But $\omega_\alpha = \sup\{\gamma_i \mid i \in I\}$, so $\omega_\alpha \leqslant \omega_\pi$, hence by 9.22, $\alpha \leqslant \pi$. It follows that α is the least upper bound of $\{\delta_i \mid i \in I\}$. ■

10.49 Lemma Let $\{\delta_i \mid i \in I\}$ be a set of ordinals, and let α be any ordinal such that $\#I < \aleph_\alpha$. Then $\sum_{i\in I} \aleph_{\delta_i} = \aleph_\alpha$ if and only if $\alpha = \sup\{\delta_i \mid i \in I\}$.

Proof. If I is finite, the result follows trivially; thus, we may assume I is infinite.

i) Suppose

$$\sum_{i\in I} \aleph_{\delta_i} = \aleph_\alpha.$$

Then for each $j \in I$,

$$\aleph_{\delta_j} \leqslant \sum_{i\in I} \aleph_{\delta_i} = \aleph_\alpha,$$

hence $\delta_j \leqslant \alpha$. Thus α is an upper bound of $\{\delta_i \mid i \in I\}$. Before going on, we note that there exists an $i \in I$ such that $\#I < \aleph_{\delta_i}$; for if $\aleph_{\delta_i} \leqslant \#I$ for every $i \in I$, then by 8.19, $\sum_{i\in I} \aleph_{\delta_i} \leqslant (\#I)(\#I) = \#I < \aleph_\alpha$, which is contrary to our assumption. Now suppose that $\delta_i < \pi$ for every $i \in I$. Then by 8.19,

$$\sum_{i\in I} \aleph_{\delta_i} \leqslant (\#I)\,\aleph_\pi.$$

But we have just seen that for some $i \in I$, $\#I < \aleph_{\delta_i}$, hence $\#I < \aleph_\pi$; thus $(\#I)\,\aleph_\pi = \aleph_\pi$. It follows that $\sum_{i\in I} \aleph_{\delta_i} \leqslant \aleph_\pi$, that is, $\aleph_\alpha \leqslant \aleph_\pi$, so by 9.22, $\alpha \leqslant \pi$. We have proved that α is the least upper bound of $\{\delta_i \mid i \in I\}$.

ii) Suppose $\alpha = \sup\{\delta_i \mid i \in I\}$. For each $i \in I$, $\delta_i \leqslant \alpha$, so by 9.22, $\aleph_{\delta_i} \leqslant \aleph_\alpha$.

Thus by 8.19,

$$\sum_{i \in I} \aleph_{\delta_i} \leqslant (\#I)\,\aleph_\alpha = \aleph_\alpha.$$

Now let $\sum_{i \in I} \aleph_{\delta_i} = \aleph_\delta$. Then for each $j \in I$,

$$\aleph_{\delta_j} \leqslant \sum_{i \in I} \aleph_{\delta_i} = \aleph_\delta,$$

hence by 9.22, $\delta_j \leqslant \delta$. But $\alpha = \sup \delta_i$, so $\alpha \leqslant \delta$, hence

$$\aleph_\alpha \leqslant \aleph_\delta = \sum_{i \in I} \aleph_{\delta_i}.$$

Thus, finally,

$$\sum_{i \in I} \aleph_{\delta_i} = \aleph_\alpha. \quad \blacksquare$$

10.50 Theorem $\aleph_\alpha$ is a regular cardinal if and only if it satisfies the following condition:

10.51 If $\{a_i \mid i \in I\}$ is any set of cardinals such that $a_i < \aleph_\alpha$ for each $i \in I$ and $\#I < \aleph_\alpha$, then $\sum_{i \in I} a_i < \aleph_\alpha$.

Proof

i) Suppose $\aleph_\alpha$ is a regular cardinal and $\{a_i \mid i \in I\}$ is a set of cardinals such that $a_i < \aleph_\alpha$ for each $i \in I$ and $\#I < \aleph_\alpha$. We may assume the a_i are all infinite cardinals, for any finite α_i among them would not affect our result; thus, we may set $a_i = \aleph_{\delta_i}$ for each $i \in I$. We now have $\aleph_{\delta_i} < \aleph_\alpha$ for each $i \in I$ and $\#I < \aleph_\alpha$, so by 8.19, $\sum_{i \in I} \aleph_{\delta_i} \leqslant \aleph_\alpha$. Now if $\sum_{i \in I} \aleph_{\delta_i} = \aleph_\alpha$, then by 10.49, $\alpha = \sup\{\delta_i \mid i \in I\}$, hence by 9.22,

$$\omega_\alpha = \sup\{\omega_{\delta_i} \mid i \in I\},$$

so by 10.42, $\{\omega_{\delta_i} \mid i \in I\}$ is a cofinal subset of ω_α. But this is impossible, for the following reason: $\#I < \aleph_\alpha$, hence

$$\#\{\omega_{\delta_i} \mid i \in I\} < \aleph_\alpha,$$

so by 10.39, $\circledcirc\{\omega_{\delta_i} \mid i \in I\} < \omega_\alpha$, which is in contradiction with the fact that ω_α is a regular ordinal.

ii) Conversely, suppose $\aleph_\alpha$ satisfies 10.51, and let $\{\gamma_i \mid i \in I\}$ be a cofinal subset of ω_α, that is,

$$\omega_\alpha = \sup\{\gamma_i \mid i \in I\}.$$

[We will assume that the γ_i are all distinct, hence $\#\{\gamma_i \mid i \in I\} = \#I$.] Set $\aleph_{\delta_i} = \#\gamma_i$ for each $i \in I$. Then, by 10.48, $\alpha = \sup\{\delta_i \mid i \in I\}$, hence by 10.49,

$$\sum_{i \in I} \aleph_{\delta_i} = \sum (\#\gamma_i) = \aleph_\alpha.$$

Thus, by 10.51, we must necessarily have $\#I \geqslant \aleph_\alpha$, so by 10.39,

$$\circledcirc\{\gamma_i \mid i \in I\} \geqslant \omega_\alpha.$$

This proves that any cofinal subset of ω_α has ordinality $\geqslant \omega_\alpha$, so ω_α is a regular ordinal. ■

10.52 Corollary If α is a nonlimit ordinal, then $\aleph_\alpha$ is a regular cardinal.

Proof. If $\alpha = \delta + 1$, then set $a_i = \aleph_\delta$ for each $i \in I$, where $\#I = \aleph_\delta$. Then by 8.17,

$$\sum_{i \in I} a_i = (\aleph_\delta)^2 = \aleph_\delta < \aleph_{\delta+1}.$$

Thus, by 10.50, $\aleph_\alpha$ is a regular cardinal. ■

10.53 Definition Let a be a cardinal number; a is called an *inaccessible* cardinal (more precisely, a *strongly inaccessible* cardinal) if

i) a is a regular cardinal, and
ii) $b < a$ and $c < a \Rightarrow b^c < a$.

ω_α is called an *inaccessible ordinal* if $\aleph_\alpha$ is an inaccessible cardinal.

In view of Theorem 10.50, an infinite cardinal number b is inaccessible if and only if it satisfies the following pair of conditions:

i) If $\{a_i \mid i \in I\}$ is a set of cardinals such that $a_i < b$ for each $i \in I$ and $\#I < b$, then $\sum_{i \in I} a_i < b$.
ii) If $a < b$ and $c < b$, then $a^c < b$.

In other words, an infinite cardinal number b is inaccessible if and only if it cannot be obtained *either as a sum of fewer than b cardinals smaller than b, or by raising a cardinal smaller than b to a power smaller than b.* This explains the use of the word "inaccessible." It can be shown, furthermore, that if b is an inaccessible cardinal, then b cannot be obtained as a product of fewer than b cardinals smaller than b (see Exercise 4, below).

We may call a nonempty set A *inaccessible* if it cannot be constructed from smaller sets by using the set-theoretical operations of union, intersection, product, or power set. To be technical, this means that

i) A is not equal to $\bigcap_{i \in I} B_i$, $\bigcup_{i \in I} B_i$, or $\prod_{i \in I} B_i$ for any set of sets $\{B_i \mid i \in I\}$, where $\{B_i \mid i \in I\} \prec A$ and $B_i \prec A$ for each $i \in I$;
ii) A is not the power set of any set B such that $B \prec A$.

For example, if we begin with the empty set and start to construct sets such as

$$\{\varnothing\}, \quad \{\varnothing, \{\{\varnothing\}\}, \{\varnothing\}\}, \quad \{\{\varnothing\}, \{\varnothing, \{\varnothing\}\}\}, \quad \text{etc.,}$$

and if we proceed to construct larger and larger sets from these by using the operations of union (including infinite union), product (including infinite product) and power set, we will never end up with an inaccessible set. Now, in view of what we have said in the preceding paragraph, it is clear that inaccessible cardinals are the cardinals of inaccessible sets.

A rather obvious question which arises in axiomatic set theory is, "Are there any inaccessible sets?" In other words, do inaccessible cardinals exist? The existence of inaccessible cardinals cannot be proved by means of the axioms we have already introduced; however, a new axiom may be added to set theory, called the *axiom for inaccessible cardinals*, asserting their existence. It has been proven in recent years that this axiom is not a consequence of the other axioms of set theory; whether it is consistent with the other axioms of set theory is still an open question.

The following definition is useful:

10.54 Definition $\aleph_\alpha$ is called a *weakly inaccessible cardinal* if

i) $\aleph_\alpha$ is a regular cardinal,

ii) α is a limit ordinal.

By Corollary 10.52, if α is a nonlimit ordinal, then $\aleph_\alpha$ is necessarily regular; thus, it is natural to ask whether there are any regular cardinals $\aleph_\alpha$ where α is a limit ordinal. Again, we cannot prove the existence of such cardinals from the usual axioms of set theory, but their existence follows immediately from the axiom for inaccessible cardinals; indeed, we have:

10.55 Theorem If a is strongly inaccessible, then a is weakly inaccessible.

Proof. Suppose $\aleph_\alpha$ is strongly inaccessible, and assume α is a nonlimit ordinal, $\alpha = \delta + 1$. By 8.16,

$$\aleph_\delta^{\aleph_\delta} = 2^{\aleph_\delta} > \aleph_\delta, \text{ hence } \aleph_\delta^{\aleph_\delta} \geqslant \aleph_\alpha$$

which is contrary to our hypothesis that $\aleph_\alpha$ is strongly inaccessible; thus α is a limit ordinal. ∎

There is another interesting connection between strong and weak inaccessibility:

10.56 Theorem Assuming the generalized continuum hypothesis to be true, if a is weakly inaccessible then a is strongly inaccessible.

Proof. Assume the generalized continuum hypothesis, let $\aleph_\alpha$ be weakly inaccessible, and suppose $\aleph_\beta < \aleph_\alpha$. By 9.22, $\beta < \alpha$, and since α is a limit ordinal, $\beta + 1 < \alpha$, so $\aleph_{\beta+1} < \aleph_\alpha$. But by the generalized continuum hypothesis, $2^{\aleph_\beta} = \aleph_{\beta+1}$, hence $2^{\aleph_\beta} < \aleph_\alpha$. It follows (see Exercise 3, Exercise Set 10.5) that $\aleph_\alpha$ is strongly inaccessible. ■

Thus, if we assume the generalized continuum hypothesis, then the notions of strongly inaccessible and weakly inaccessible are equivalent.

EXERCISES 10.5

1. Prove 10.37 and 10.38.
2. Prove 10.39.
3. Prove that a is strongly inaccessible if and only if
   a) a is regular, and
   b) $\forall b < a, 2^b < a$. [See Exercise 3, Exercise Set 8.4.]
4. Prove that if b is an inaccessible cardinal, then b satisfies the following condition:

   If $\{a_i \mid i \in I\}$ is a set of cardinals such that $a_i < b$ for each $i \in I$ and $\#I < b$, then $\prod_{i \in I} a_i < b$.
5. Prove that if an infinite cardinal b satisfies the condition in Exercise 4, then b is an inaccessible cardinal.

Bibliography

1. Bourbaki, N., *Théorie des Ensembles,* Paris, Hermann, 1963.
2. Fraenkel, A. A., *Set Theory and Logic,* Reading, Mass., Addison-Wesley, 1966.
3. Halmos, P., *Naive Set Theory*, Princeton, Van Nostrand, 1960.
4. Kamke, E., *Theory of Sets,* New York, Dover, 1950.
5. Monk, J. D., *Introduction to Set Theory,* New York, McGraw-Hill, 1969.
6. Quine, W. V., *Mathematical Logic,* Cambridge, Mass., Harvard University Press, 1951.
7. Rubin, J. E., *Set Theory for the Mathematician,* San Francisco, Holden-Day, 1966.
8. Slupecki, J. and L. Borkowski, *Elements of Mathematical Logic and Set Theory,* Oxford, Pergamon Press, 1967.
9. Suppes, P., *Introduction to Logic,* Princeton, Van Nostrand, 1957.
10. Suppes, P., *Axiomatic Set Theory,* Princeton, Van Nostrand, 1960.

Index